FOURTH EDITION

Atlas of Skeletal Muscles

Robert J. Stone
Suffolk County Community College

Judith A. Stone
Suffolk County Community College

Boston   Burr Ridge, IL   Dubuque, IA   Madison, WI   New York   San Francisco   St. Louis
Bangkok   Bogotá   Caracas   Kuala Lumpur   Lisbon   London   Madrid   Mexico City
Milan   Montreal   New Delhi   Santiago   Seoul   Singapore   Sydney   Taipei   Toronto

McGraw-Hill Higher Education 🪐

A Division of The McGraw-Hill Companies

ATLAS OF SKELETAL MUSCLES, FOURTH EDITION

Published by McGraw-Hill, a business unit of The McGraw-Hill Companies, Inc., 1221 Avenue of the Americas, New York, NY 10020. Copyright © 2003, 2000, 1997 by The McGraw-Hill Companies, Inc. All rights reserved. No part of this publication may be reproduced or distributed in any form or by any means, or stored in a database or retrieval system, without the prior written consent of The McGraw-Hill Companies, Inc., including, but not limited to, in any network or other electronic storage or transmission, or broadcast for distance learning.

Some ancillaries, including electronic and print components, may not be available to customers outside the United States.

♲ This book is printed on recycled, acid-free paper containing 10% postconsumer waste.

International   2 3 4 5 6 7 8 9 0 QPD/QPD 0 9 8 7 6 5 4 3
Domestic   3 4 5 6 7 8 9 0 QPD/QPD 0 9 8 7 6 5 4 3

ISBN 0–07–235080–6
ISBN 0–07–119901–2 (ISE)

Publisher: *Martin J. Lange*
Developmental editor: *Kristine A. Queck*
Senior development manager: *Kristine Tibbetts*
Senior marketing manager: *Michelle Watnick*
Project manager: *Joyce Watters*
Senior production supervisor: *Sandy Ludovissy*
Coordinator of freelance design: *Michelle D. Whitaker*
Cover/interior designer: *Elise Lansdon*
Cover illustrations: *Robert J. Stone;* Inset image: *PhotoDisc, Fitness & Well-Being, #67018*
Lead photo research coordinator: *Carrie K. Burger*
Compositor: *Shepherd, Inc.*
Typeface: *10/12 Frutiger Roman*
Printer: *Quebecor World Dubuque, IA*

Library of Congress Cataloging-in-Publication Data

Stone, Robert J.
    Atlas of skeletal muscles / Robert J. Stone, Judith A. Stone. — 4th ed.
       p.   cm.
    Includes index.
    ISBN 0–07–235080–6
    1. Striated muscles—Atlases. I. Stone, Judith A. II. Title.

QM151 .S76 2003
611'.73'0222—dc21   2002025082
                                                     CIP

INTERNATIONAL EDITION ISBN 0–07–119901–2
Copyright © 2003. Exclusive rights by The McGraw-Hill Companies, Inc., for manufacture and export. This book cannot be re-exported from the country to which it is sold by McGraw-Hill. The International Edition is not available in North America.

www.mhhe.com

Dedication

To Karen, Andrew, and Laura, helping us more and more as the years go by.

Contents

CHAPTER FOUR

Muscles of the Neck 49

CHAPTER FIVE

Muscles of the Trunk 71

CHAPTER SIX

Muscles of the Shoulder and Arm 97

CHAPTER SEVEN

Muscles of the Forearm and Hand 121

CHAPTER EIGHT

Muscles of the Hip and Thigh 155

CHAPTER NINE

Muscles of the Leg and Foot 185

Preface

This book is a study guide and reference for the anatomy and actions of human skeletal muscles. It is designed for use by students of anatomy, physical education, and health-related fields. It also serves as a compact reference for the practicing professional.

The first chapter presents labeled line drawings of the skeleton, which include all structures that are used in describing origins and insertions in the later chapters. A master numbering system is used so that each structure is labeled with the same number in all drawings.

The second chapter describes the various movements of the body.

In chapters 3 through 9, the origin, insertion, action, and innervation of the skeletal muscles are described, and each muscle is presented on a separate page with a line drawing.

The spinal cord levels of the nerve fibers that innervate each muscle are included in parentheses after the name of each nerve.

Labeled drawings of major muscle groups are presented throughout chapters 3 through 9. Notes and relationships among muscles have been included on many pages.

The drawings include the following important features:

1. Bones and cartilage containing muscle attachment sites are shaded.
2. Adjacent structures are shown.
3. Muscle fibers are drawn by direction.
4. Muscle fibers are shown on the undersurface of bone and cartilage as dashed lines.
5. Tendons and aponeuroses are shown.
6. Labeled muscle groups are included.

These features aid in visual orientation and understanding of the action of the muscles. We have noticed that many students find it useful to color the illustrations.

Notes have been included on many pages to show how muscles are used. Relationships among many of the muscles have also been indicated where appropriate.

Some users of previous editions have advised that some of the smaller muscles should be enlarged and shown with less skeletal background. We have purposely standardized the skeletal views to allow an appreciation of the relative sizes and positions of the muscles. Since skeletal muscles, at the gross level, are relatively simple anatomical structures, very little additional information would be included by enlargement, and many comparative relationships would be lost.

Since our primary goal is to describe the muscles moving the skeleton, we have not described the muscles of the perineum, eye, tympanic cavity, tongue, larynx, pharynx, or palate.

We extend our appreciation to Mr. George Boykin, who was for many years the jolly proprietor of the gross anatomy laboratories at the Health Sciences Center of the State University of New York at Stony Brook, for his help and encouragement. We also thank Mr. Vincent Verdisco and Ms. Diane Chandler for their technical advice with the artwork and Ms. Katherine Juner for her secretarial services.

We would also like to thank the many reviewers who have made helpful suggestions for improving past editions of this atlas, as well as Kelly Johnson, University of Kansas; John H. Langdon, University of Indianapolis; Jeff Seegmiller, Illinois State University; and Curt Walker, Dixie State College for their input on the fourth edition.

Robert J. Stone

Judith A. Stone

CHAPTER ONE

The Skeleton

SKULL—LATERAL VIEW

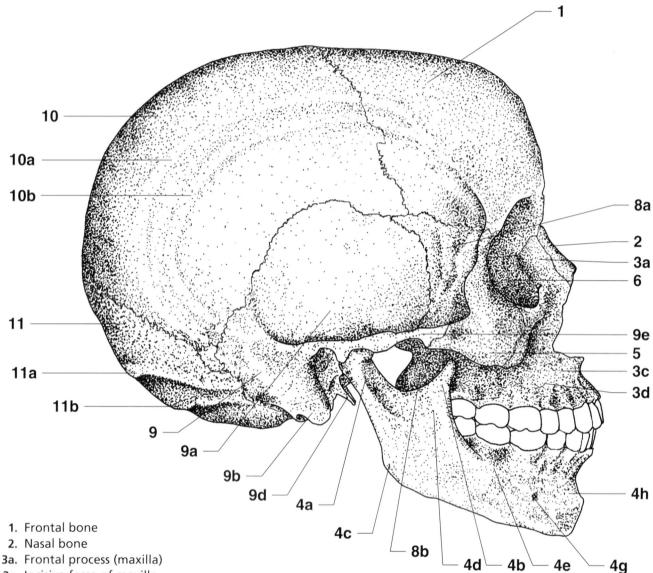

1. Frontal bone
2. Nasal bone
3a. Frontal process (maxilla)
3c. Incisive fossa of maxilla
3d. Canine fossa (maxilla)
4a. Neck of condyle (mandible)
4b. Coronoid process (mandible)
4c. Angle of the mandible
4d. Ramus (mandible)
4e. Oblique line (mandible)
4g. Mental foramen (mandible)
4h. Incisive fossa of mandible
5. Zygomatic bone
6. Lacrimal bone
8a. Greater wing of sphenoid bone
8b. Lateral pterygoid plate
9. Temporal bone

9a. Temporal fossa
9b. Mastoid process (temporal bone)
9d. Styloid process (temporal bone)
9e. Zygomatic process (temporal bone)
10. Parietal bone
10a. Superior temporal line
10b. Inferior temporal line
11. Occipital bone
11a. Superior nuchal line (occipital bone)
11b. Inferior nuchal line (occipital bone)

Note: The zygomatic arch is formed by the zygomatic process of the temporal bone meeting the zygomatic bone.

ATLAS AND AXIS

TEMPOROMANDIBULAR JOINT

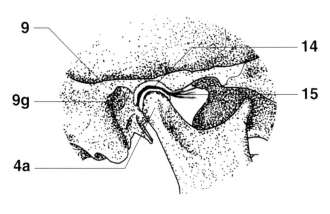

22a. Transverse process of atlas
22b. Superior articular facet of atlas
22c. Vertebral foramina of atlas and axis
22d. Anterior tubercle of atlas
23b. Dens of axis
23c. Transverse foramina of atlas and axis
23d. Lamina
23e. Spinous process of axis

4a. Neck of condyle (mandible)
 9. Temporal bone
9g. External acoustic meatus (canal)
14. Articular disk
15. Lateral pterygoid muscle

SKULL—ANTERIOR VIEW

SKULL—INFERIOR (BASAL) VIEW

1. Frontal bone
1c. Superciliary arch (frontal bone)
2. Nasal bone
3. Maxilla
3c. Incisive fossa of maxilla
3e. Alveolar border of maxilla
4. Mandible
4i. Symphysis of mandible
5. Zygomatic bone
6. Lacrimal bone
7a. Nasal septum (ethmoid bone)
8a. Greater wing of sphenoid bone
8c. Lesser wing of sphenoid bone

5a. Zygomatic arch
8b. Lateral pterygoid plate
9b. Mastoid process (temporal bone)
9f. Mandibular process (temporal bone)
11a. Superior nuchal line (occipital bone)
11b. Inferior nuchal line (occipital bone)
11c. External occipital protuberance (occipital bone)
11d. Occipital condyle (occipital bone)
11e. External occipital crest (occipital bone)

SKULL TO HUMERUS—LATERAL VIEW

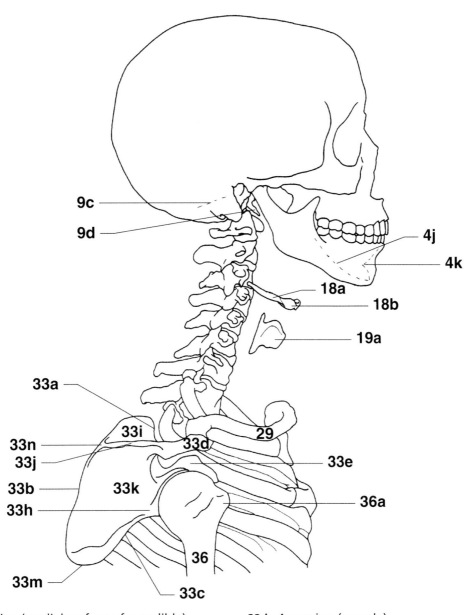

4j. Mylohyoid line (medial surface of mandible)
4k. Inferior mental spine (inner surface of mandible)
9c. Mastoid notch (medial surface of temporal bone)
9d. Styloid process (temporal bone)
18a. Greater cornu of hyoid
18b. Body of hyoid
19a. Lamina of thyroid cartilage
29. Clavicle
33a. Superior border of scapula
33b. Vertebral (medial) border of scapula
33c. Axillary (lateral) border of scapula

33d. Acromion (scapula)
33e. Coracoid process (scapula)
33h. Infraglenoid tubercle (scapula)
33i. Supraspinous fossa (scapula)
33j. Crest of spine (scapula)
33k. Infraspinous fossa (scapula)
33m. Inferior angle of scapula
33n. Root of spine (scapula)
36. Humerus
36a. Greater tuberosity (tubercle) of humerus

SKULL TO STERNUM—ANTERIOR VIEW *(Mandible and Maxilla Removed)*

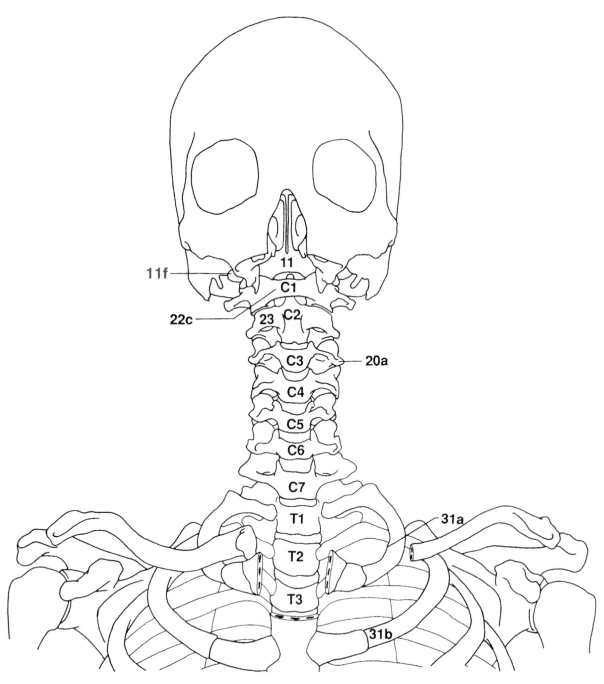

11. Occipital bone
11f. Jugular process of occipital bone
20a. Transverse process of vertebra
22c. Anterior arch of atlas

23. Axis
31a. Scalene tubercle of first rib
31b. Second rib

RIB CAGE, PECTORAL GIRDLE, UPPER ARM—
ANTERIOR VIEW *(Ribs Partially Removed, Right Arm Disarticulated)*

29. Clavicle
30a. Manubrium (sternum)
30b. Body (sternum)
30c. Xiphoid process (sternum)
31c. Tubercle of rib
31d. Angle of rib
32. Costal cartilage
33d. Acromion (scapula)
33e. Coracoid process (scapula)
33f. Supraglenoid tubercle (scapula)
33l. Subscapular fossa (scapula)

36a. Greater tuberosity (tubercle) of humerus
36b. Lesser tuberosity of humerus
36c. Intertubercular (bicipital) groove (humerus)
36d. Deltoid tuberosity (humerus)
37a. Radial tuberosity (radius)
38a. Ulnar tuberosity (ulna)
38b. Coronoid process (ulna)
92. Lateral lumbocostal arch (lateral arcuate ligament)*
93. Medial lumbocostal arch (medial arcuate ligament)

*See p. 89.

SKELETON—POSTERIOR VIEW *(Enlargement of Lumbar Vertebrae)*

20a. Transverse process of vertebra
20b. Spinous process of vertebra
20c. Mammillary process of vertebra
20d. Accessory process of vertebra
24. Ligamentum nuchae
25. Supraspinous ligament
26a. Lateral sacral crest
26b. Sacral foramina
26c. Medial sacral crest
29. Clavicle
31c. Tubercle of rib
31d. Angle of rib
33. Scapula
33d. Acromion (scapula)
36. Humerus
37. Radius
38. Ulna
38c. Olecranon process (ulna)
52a. Posterior superior iliac spine
52b. Iliac crest

RIGHT ARM—POSTERIOR VIEW

34. Radial collateral ligament
35. Annular ligament
36. Humerus
36e. Lateral supracondylar ridge (humerus)
36f. Lateral epicondyle (humerus)
37. Radius
37c. Dorsal tubercle (radius)
37d. Styloid process (radius)
38. Ulna
38c. Olecranon process (ulna)
39. Interosseous membrane
40. Scaphoid (navicular)
41. Lunate
43. Trapezium
44. Trapezoid
45. Capitate
46. Hamate
47. Triquetrum
48. Metacarpals
49a. Proximal (first) phalanges
49b. Middle (second) phalanges
49c. Distal (third) phalanges

RIGHT HAND— ANTERIOR VIEW

RIGHT ARM— ANTERIOR VIEW

34. Radial collateral ligament
35. Annular ligament
36. Humerus
36g. Medial supracondylar ridge (humerus)
36h. Medial epicondyle (humerus)
37. Radius
37b. Pronator tuberosity (radius)
38. Ulna
38b. Coronoid process (ulna)
38c. Olecranon process (ulna)
38d. Supinator crest (ulna)
39. Interosseous membrane
40a. Tubercle of scaphoid (navicular)
41. Lunate
42. Pisiform
43a. Tubercle of trapezium
44. Trapezoid
45. Capitate
46a. Hook of hamate
47. Triquetrum
48. Metacarpals
49a. Proximal (first) phalanges
49b. Middle (second) phalanges
49c. Distal (third) phalanges
50. Flexor retinaculum
51. Palmar aponeurosis

LUMBAR AND PELVIC REGION—ANTERIOR VIEW

20a. Transverse process of vertebra
20e. Body of vertebra
21. Intervertebral disk
26d. Second sacral vertebra
26e. Third sacral vertebra
26f. Fourth sacral vertebra
26g. Ala of sacrum
52e. Anterior superior iliac spine
52f. Anterior inferior iliac spine
52g. Iliac fossa
52h. Arcuate line (ilium)
52i. Iliopectineal eminence (ilium)
53a. Ramus of ischium
54a. Superior ramus of pubis
54b. Inferior ramus of pubis
54c. Pubic crest
54d. Pubic symphysis
54e. Pubic tubercle
55. Greater sciatic notch
56. Sacrotuberous ligament
57. Obturator foramen
59. Inguinal ligament
60. Superior pubic ligament
61. Iliolumbar ligament
61. Rectus sheath
65. Linea alba
68. Femur
68a. Greater trochanter (femur)
68b. Trochanteric fossa (femur)
68c. Lesser trochanter (femur)

PELVIC GIRDLE TO KNEE— LATERAL VIEW

THORACIC TO PELVIC REGION— LATERAL VIEW
(Arm and Leg Removed)

52b. Iliac crest
52c. Iliac tubercle
52e. Anterior superior iliac spine
52f. Anterior inferior iliac spine
67. Iliotibial tract
68. Femur
70. Patella
72. Synovial membrane of knee joint
74c. Lateral condyle of tibia

32. Costal cartilage
33g. Glenoid cavity (scapula)
52d. Anterior iliac crest
59. Inguinal ligament
63. Acetabulum
64. Thoracolumbar fascia
66. Abdominal aponeurosis

PELVIC GIRDLE—POSTERIOR VIEW

52. Ilium
53a. Ramus of ischium
53b. Ischial tuberosity
53c. Spine of ischium
54g. Body of pubis
57. Obturator foramen
58. Obturator membrane
68a. Greater trochanter (femur)
68c. Lesser trochanter (femur)
68d. Intertrochanteric crest (femur)

PELVIC GIRDLE AND UPPER LEG—THREE-QUARTER POSTERIOR VIEW

26. Sacrum
27. Aponeurosis of erector spinae
28. Coccyx
52b. Iliac crest
52j. Posterior gluteal line (ilium)
52k. Middle (anterior) gluteal line (ilium)
52l. Inferior gluteal line (ilium)
53d. Lesser sciatic notch
55. Greater sciatic notch
56. Sacrotuberous ligament
68a. Greater trochanter (femur)
68g. Gluteal tuberosity (femur)
68h. Linea aspera (femur)

PELVIC GIRDLE TO LEG—ANTERIOR VIEW

52. Ilium
52e. Anterior superior iliac spine
52f. Anterior inferior iliac spine
54b. Inferior ramus of pubis
54f. Pectineal line (pubis)
54g. Body of pubis
68. Femur
68a. Greater trochanter (femur)
68c. Lesser trochanter (femur)
68e. Intertrochanteric line (femur)
68i. Lateral supracondylar line (femur)
68j. Medial supracondylar line (femur)
69. Quadriceps tendon
70. Patella
71. Patellar ligament
74. Tibia
74a. Tuberosity of tibia

PELVIC GIRDLE TO LEG—POSTERIOR VIEW

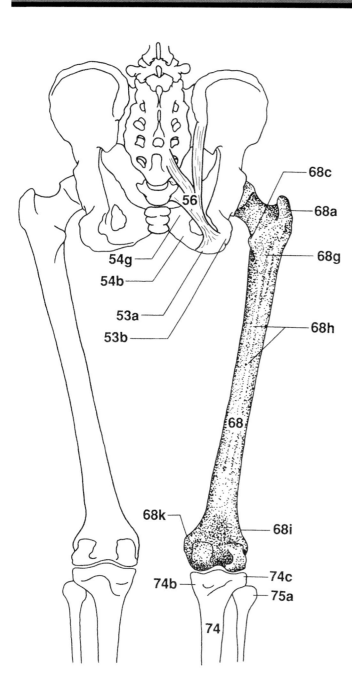

53a. Ramus of ischium
53b. Ischial tuberosity
54b. Inferior ramus of pubis
54g. Body of pubis
56. Sacrotuberous ligament
68. Femur
68a. Greater trochanter (femur)
68c. Lesser trochanter (femur)
68g. Gluteal tuberosity (femur)
68h. Linea aspera (femur)
68i. Lateral supracondylar line (femur)
68k. Adductor tubercle (femur)
74. Tibia
74b. Medial condyle of tibia
74c. Lateral condyle of tibia
75a. Head of fibula

RIGHT FOOT— ANTEROLATERAL VIEW

RIGHT LEG— ANTEROLATERAL VIEW

68. Femur
74. Tibia
74c. Lateral condyle of tibia
75. Fibula
76. Interosseous membrane
77. Lateral talocalcaneal ligament
78. Inferior extensor retinaculum
83. Talus
84. Calcaneus
85. Navicular
86. Medial cuneiform
87. Intermediate cuneiform
88. Lateral cuneiform
89. Cuboid
90. Metatarsal bones
90a. First metatarsal
90b. Second metatarsal
90c. Third metatarsal
90d. Fourth metatarsal
90e. Fifth metatarsal
91a. Proximal phalanges
91b. Middle phalanges
91c. Distal phalanges

RIGHT LEG—POSTERIOR VIEW

68. Femur
68i. Lateral supracondylar line (femur)
68l. Popliteal surface (femur)
68m. Medial condyle (femur)
68n. Lateral condyle (femur)
73. Oblique popliteal ligament
74. Tibia
74d. Soleal line (tibia)
75. Fibula
76. Interosseous membrane
84. Calcaneus
84a. Sustentaculum tali of calcaneus
85a. Tuberosity of navicular
86. Medial cuneiform
87. Intermediate cuneiform
88. Lateral cuneiform
89. Cuboid

RIGHT FOOT—PLANTAR VIEW

74e. Medial malleolus of tibia
79. Flexor retinaculum
80. Plantar aponeurosis
81. Plantar metatarsophalangeal ligaments
82. Transverse metatarsal ligaments
83. Talus
84a. Sustentaculum tali of calcaneus
84b. Tuberosity of calcaneus
84c. Medial border of calcaneus
84d. Lateral border of calcaneus

85a. Tuberosity of navicular
86. Medial cuneiform
87. Intermediate cuneiform
88. Lateral cuneiform
89. Cuboid
90. Metatarsal bones
91a. Proximal phalanges
91b. Middle phalanges
91c. Distal phalanges

CHAPTER TWO

Movements of the Body

Anatomical position—A subject in the anatomical position is standing erect with the head, eyes, and toes facing forward and the arms hanging straight at the sides with the palms of the hands facing forward.

FIGURE 2.1
Median or midsagittal plane—Passes vertically through the body from anterior (front) to posterior (back). It divides the body into right and left sides. Other sagittal planes are parallel to this plane.

FIGURE 2.2
Coronal (frontal) planes—Pass vertically through the body from side to side. They divide the body from front to back.

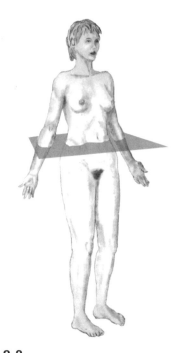

FIGURE 2.3
Transverse planes (cross sections)—Pass horizontally through the body parallel to the ground.

F I G U R E 2.4
Flexion—The left arm, forearm, and right thigh are drawn forward in sagittal planes. The right knee is also flexed.
Extension—The left thigh and knee are extended.
Hyperextension—The right arm is hyperextended at the shoulder.

F I G U R E 2.5
Lateral flexion—The torso (or head) bends laterally in the coronal plane.

F I G U R E 2.6
Abduction—The right leg moves laterally in the coronal plane.
Adduction—The leg is returned medially in the coronal plane.

Note: If the foot is fixed, this motion results in tilting the pelvis upward on the opposite side.

F I G U R E 2.7
Medial rotation—The anterior of the arm (or thigh) is moved toward the median plane.
Lateral rotation—The anterior of the arm (or thigh) is moved away from the median plane.

MOVEMENTS OF THE SCAPULA

F I G U R E 2.8
Elevation—The right
scapula of this figure is
drawn superiorly against
the resistance of the rock.

F I G U R E 2.9
Depression—The right
scapula of this figure is
pushing the arm inferiorly.

F I G U R E 2.10
Protraction—The scapula pushes the arm forward
in a sagittal plane.

F I G U R E 2.11
Retraction—The scapula is pulled back from protraction
in a sagittal plane. Since the scapula slides around the
ribs toward the median plane, it becomes adduction.

F I G U R E 2.12
Rotation—For abduction of the arm to continue above
the height of the shoulder, the scapula must rotate on
its axis so that the glenoid fossa turns upward.

MOVEMENTS OF THE HAND AND FOREARM

FIGURE **2.13**
Pronation—The forearm is rotated away from the anatomical position so that the palm turns medially, then posteriorly. If the forearm is flexed at the elbow, then the palm turns inferiorly.
Supination—The forearm is rotated so that the palm turns anteriorly (or superiorly if the forearm is flexed). Also see figs. 2.19 and 2.20.

FIGURE **2.15**
Radial flexion (abduction)—The hand, at the wrist, is drawn away from the body in a coronal plane.
Ulnar flexion (adduction)—The hand, at the wrist, is drawn toward the body in a coronal plane.

FIGURE **2.14**
Abduction—The fingers are moved away from the midline of the hand.

FIGURE **2.16**
Adduction—The fingers are moved toward the midline of the hand.

FIGURE **2.17**
Opposition—The thumb is rotated so its anterior pad can touch the anterior pads of the four fingers.

MOVEMENTS OF THE FOOT

FIGURE 2.18
Dorsiflexion—The ankle flexes, moving the foot superiorly.
Plantar Flexion—The ankle extends, moving the foot inferiorly.

FIGURE 2.19
Eversion—The front of the foot moves laterally away from the midline (abduction), and the sole turns outward. This is also called pronation.

FIGURE 2.20
Inversion—The front of the foot moves medially toward the midline (adduction), and the sole turns inward. This is also called supination.

CHAPTER THREE

Muscles of the Face and Head

EPICRANIUS

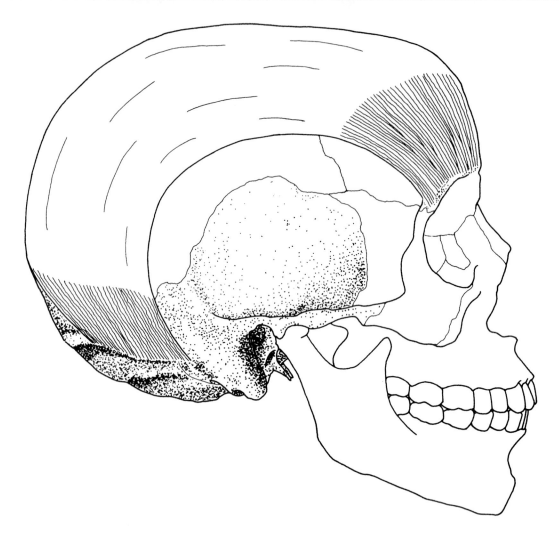

Skull—lateral view

Occipital belly (*occipitalis*)

■ **Origin**
Lateral two-thirds of superior nuchal line of occipital bone, mastoid process of temporal bone

■ **Insertion**
Galea aponeurotica (an intermediate tendon leading to frontal belly)

■ **Action**
Draws back scalp, aids frontal belly to wrinkle forehead and raise eyebrows

■ **Nerve**
Posterior auricular branch of facial nerve

Frontal belly (*frontalis*)

■ **Origin**
Galea aponeurotica

■ **Insertion**
Fascia of facial muscles and skin above nose and eyes

■ **Action**
Draws back scalp, wrinkles forehead, raises eyebrows

■ **Nerve**
Temporal branches of facial nerve

TEMPOROPARIETALIS

Skull—lateral view

■ **Origin**
Fascia over ear

■ **Insertion**
Lateral border of galea aponeurotica

■ **Action**
Raises ears, tightens scalp

■ **Nerve**
Temporal branch of facial nerve

AURICULARIS ANTERIOR, SUPERIOR, POSTERIOR

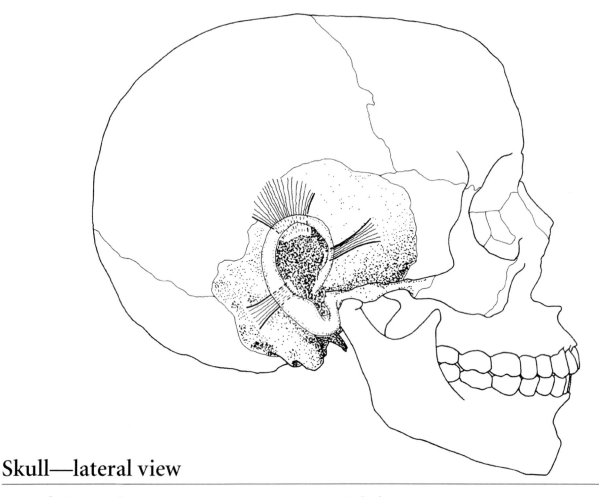

Skull—lateral view

Auricularis anterior

■ **Origin**
Fascia in temporal region

■ **Insertion**
Anterior to helix of ear

■ **Action**
Draws ear forward in some individuals, moves scalp*

■ **Nerve**
Temporal branch of facial nerve

Auricularis superior

■ **Origin**
Fascia in temporal region

■ **Insertion**
Superior part of ear

■ **Action**
Draws ear upward in some individuals, moves scalp*

■ **Nerve**
Temporal branch of facial nerve

Auricularis posterior

■ **Origin**
Mastoid area of temporal bone

■ **Insertion**
Posterior part of ear

■ **Action**
Draws ear upward in some individuals*

■ **Nerve**
Posterior auricular branch of facial nerve

*This muscle is nonfunctional in most people.

ORBICULARIS OCULI

ORBITAL AND PALPEBRAL PARTS

LACRIMAL PART

Skull—three-quarter anterior view

Orbital part

■ **Origin**

Frontal bone, maxilla (medial margin of orbit)

■ **Insertion**

Continues around orbit and returns to origin

■ **Action**

Strong closure of eyelids

■ **Nerve**

Temporal and zygomatic branches of facial nerve

Palpebral part *(in eyelids)*

■ **Origin**

Medial palpebral ligament

■ **Insertion**

Lateral palpebral ligament into zygomatic bone

■ **Action**

Gentle closure of eyelids

■ **Nerve**

Temporal and zygomatic branches of facial nerve

Lacrimal part *(behind medial palpebral ligament and lacrimal sac)*

■ **Origin**

Lacrimal bone

■ **Insertion**

Lateral palpebral raphe

■ **Action**

Draws lacrimal canals onto surface of eye

■ **Nerve**

Temporal and zygomatic branches of facial nerve

LEVATOR PALPEBRAE SUPERIORIS

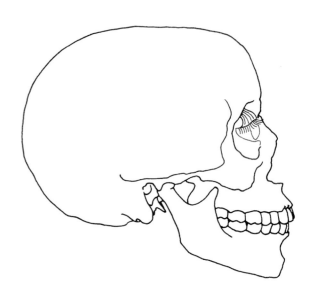

Skull—three-quarter anterior view

Skull—lateral view

■ Origin
Inferior surface of lesser wing of sphenoid

■ Insertion
Skin of upper eyelid

■ Action
Raises upper eyelid

■ Nerve
Oculomotor nerve

CORRUGATOR SUPERCILII

Skull—anterior view

■ **Origin**
Medial end of superciliary arch

■ **Insertion**
Deep surface of skin under medial portion of eyebrows

■ **Action**
Draws eyebrows downward and medially

■ **Nerve**
Temporal branch of facial nerve

PROCERUS

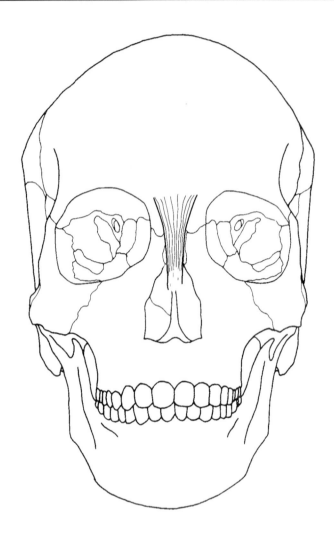

Skull—anterior view

■ Origin
Fascia over nasal bone and lateral nasal cartilage

■ Insertion
Skin between eyebrows

■ Action
Draws down medial part of eyebrows, wrinkles nose

■ Nerve
Buccal branches of facial nerve

NASALIS

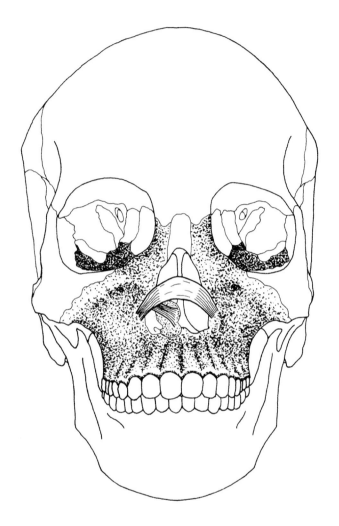

Skull—anterior view

Transverse part

■ **Origin**
Middle of maxilla

■ **Insertion**
Muscle of opposite side over bridge of nose

Alar part

■ **Origin**
Greater alar cartilage, skin on nose

■ **Insertion**
Skin at point of nose

■ **Action**
Both parts maintain opening of external nares during forceful inspiration

■ **Nerve**
Buccal branches of facial nerve

DEPRESSOR SEPTI

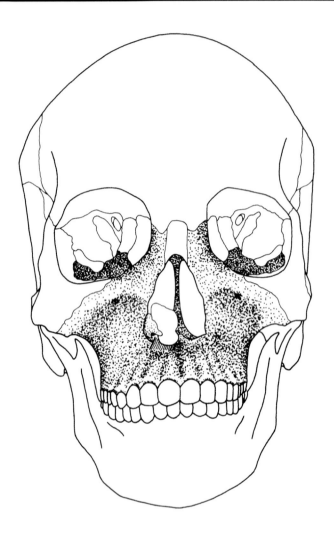

Skull—anterior view

■ **Origin**
Incisive fossa of maxilla

■ **Insertion**
Nasal septum and ala

■ **Action**
Constricts nares

■ **Nerve**
Buccal branches of facial nerve

ORBICULARIS ORIS

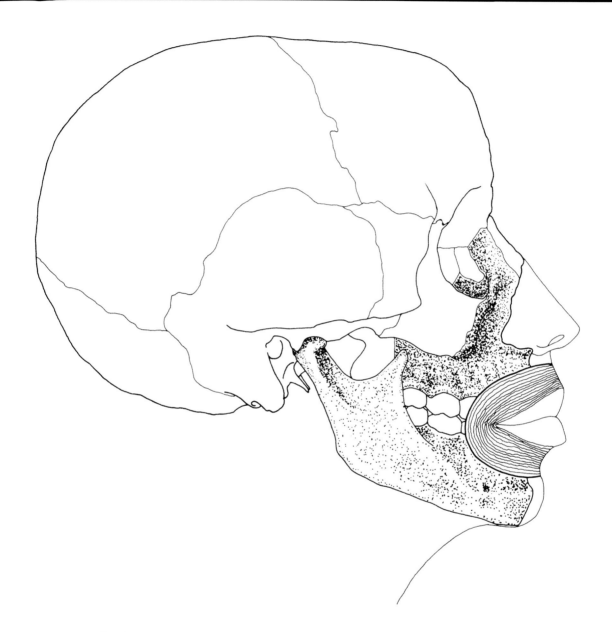

Skull—lateral view

■ Origin
Lateral band—alveolar border of maxilla

Medial band—septum of nose

Inferior portion—lateral to midline of mandible

■ Insertion
Becomes continuous with other muscles at angle of mouth

■ Action
Closure and protrusion of lips

■ Nerve
Buccal and mandibular branches of facial nerve

LEVATOR LABII SUPERIORIS

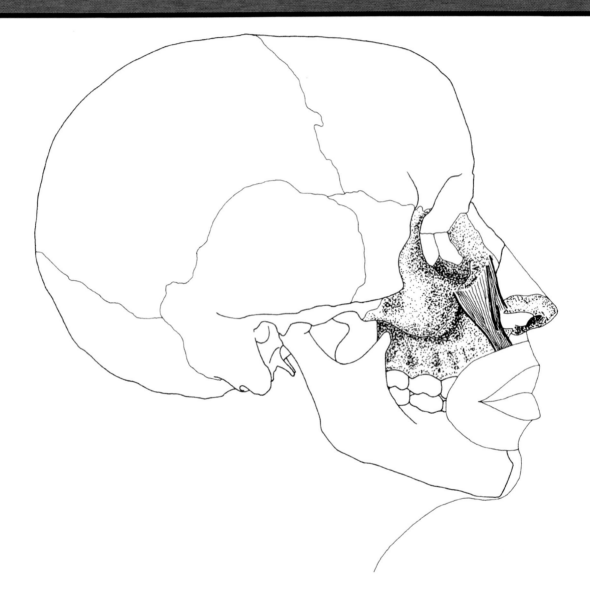

Skull—lateral view

Angular head

■ **Origin**
Frontal process of maxilla and zygomatic bone

■ **Insertion**
Greater alar cartilage and skin of nose, upper lip

■ **Action**
Elevates upper lip, dilates nares, forms nasolabial furrow

■ **Nerve**
Buccal branches of facial nerve

Infraorbital head

■ **Origin**
Lower margin of orbit

■ **Insertion**
Muscles of upper lip

■ **Action**
Elevates upper lip

■ **Nerve**
Buccal branches of facial nerve

LEVATOR ANGULI ORIS

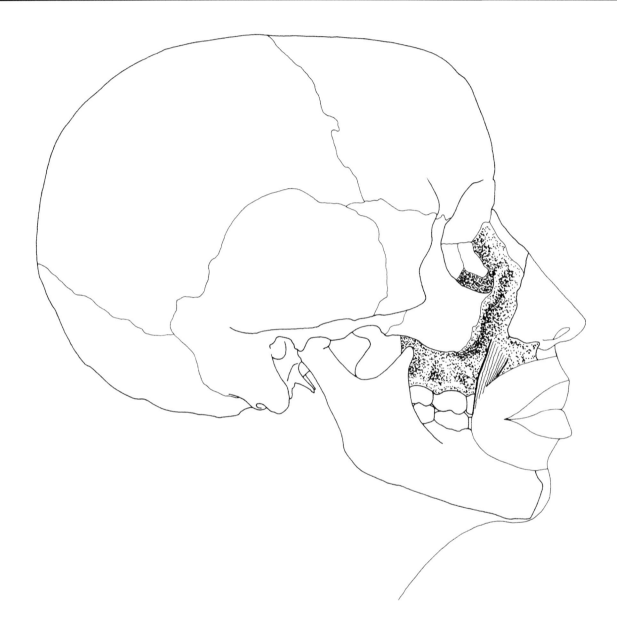

Skull—lateral view

■ **Origin**
Canine fossa of maxilla

■ **Insertion**
Angle of mouth

■ **Action**
Elevates corner (angle) of mouth

■ **Nerve**
Buccal branches of facial nerve

ZYGOMATICUS MAJOR

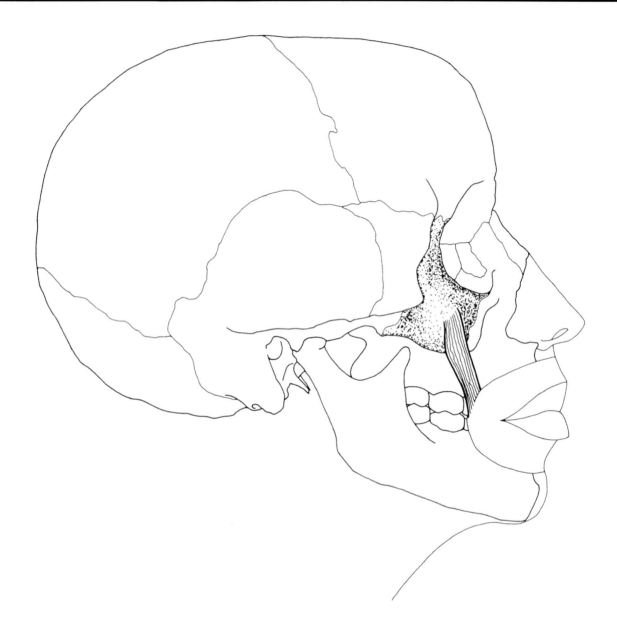

Skull—lateral view

■ **Origin**
Zygomatic bone

■ **Insertion**
Angle of mouth

■ **Action**
Draws angle of mouth upward and backward (laughing)

■ **Nerve**
Buccal branches of facial nerve

ZYGOMATICUS MINOR

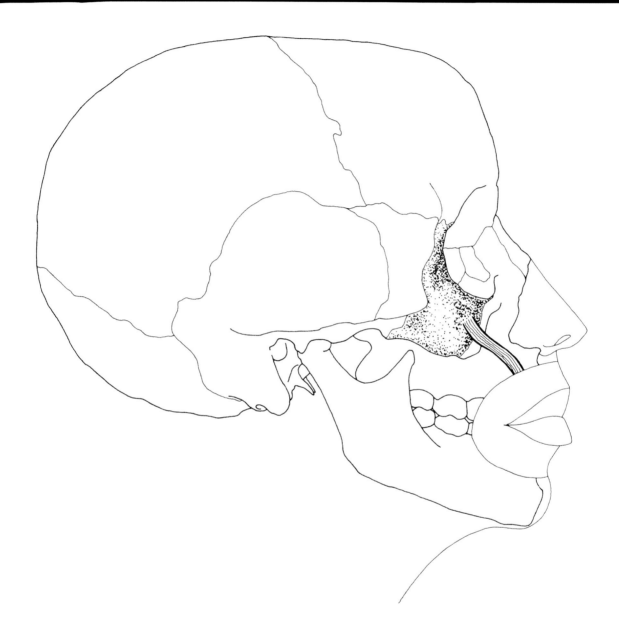

Skull—lateral view

■ **Origin**
Zygomatic bone

■ **Insertion**
Upper lip lateral to levator labii superioris

■ **Action**
Forms nasolabial furrow

■ **Nerve**
Buccal branches of facial nerve

RISORIUS

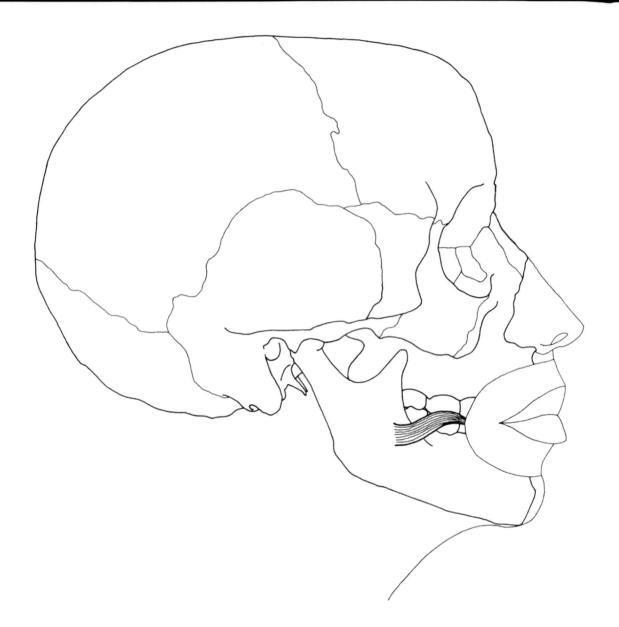

Skull—lateral view

■ **Origin**
Fascia over masseter

■ **Insertion**
Skin at angle of mouth

■ **Action**
Retracts angle of mouth, as in grinning

■ **Nerve**
Buccal branches of facial nerve

DEPRESSOR LABII INFERIORIS

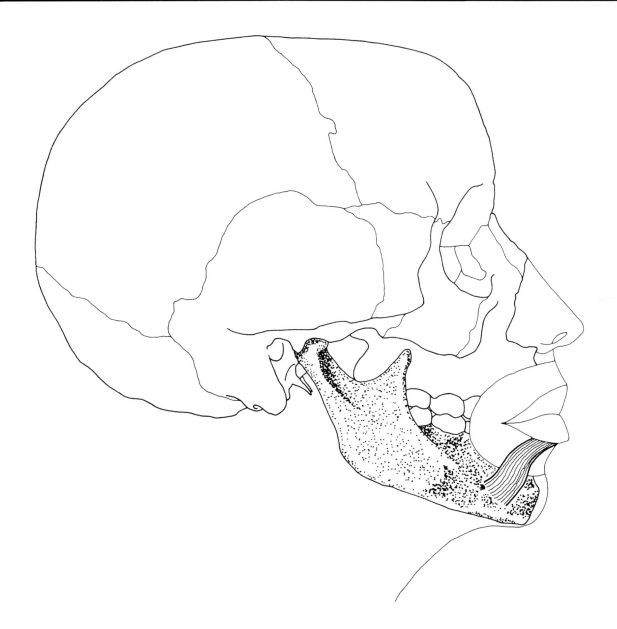

Skull—lateral view

■ **Origin**
Mandible, between symphysis and mental foramen

■ **Insertion**
Skin of lower lip

■ **Action**
Draws lower lip downward and laterally

■ **Nerve**
Mandibular branch of facial nerve

DEPRESSOR ANGULI ORIS

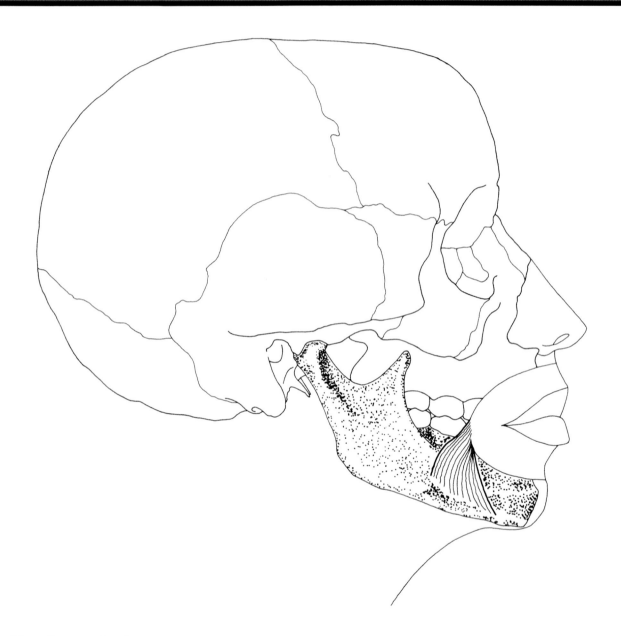

Skull—lateral view

■ **Origin**
Oblique line of the mandible

■ **Insertion**
Angle of the mouth

■ **Action**
Depresses angle of mouth, as in frowning

■ **Nerve**
Mandibular branch of facial nerve

MENTALIS

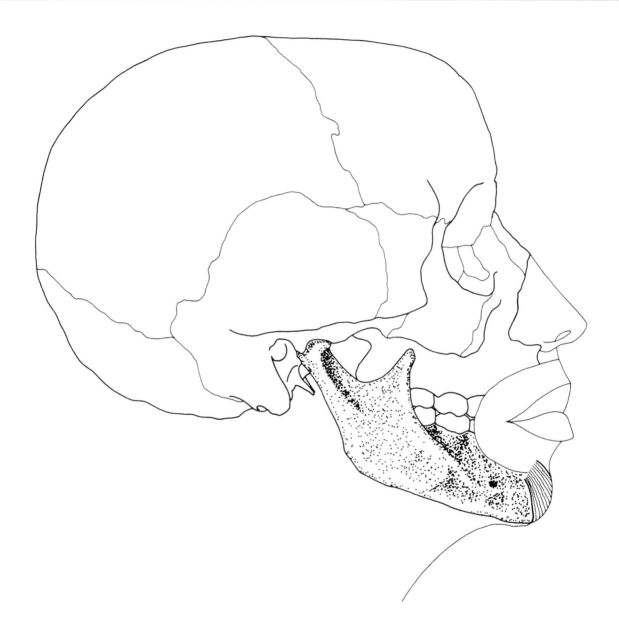

Skull—lateral view

■ **Origin**
Incisive fossa of mandible

■ **Insertion**
Skin of chin

■ **Action**
Raises and protrudes lower lip, wrinkles skin of chin

■ **Nerve**
Mandibular branch of facial nerve

BUCCINATOR

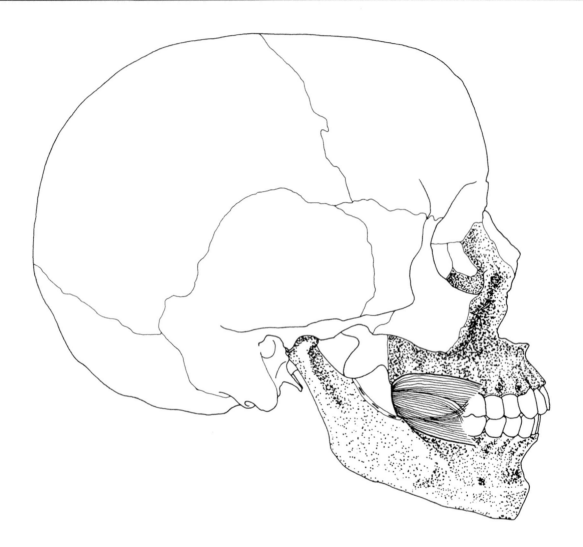

Skull—lateral view

■ **Origin**
Outer surface of alveolar processes of maxilla and mandible over molars and along pterygomandibular raphe

■ **Insertion**
Deep part of muscles of lips

■ **Action**
Compresses cheek

■ **Nerve**
Buccal branches of facial nerve

TEMPORALIS

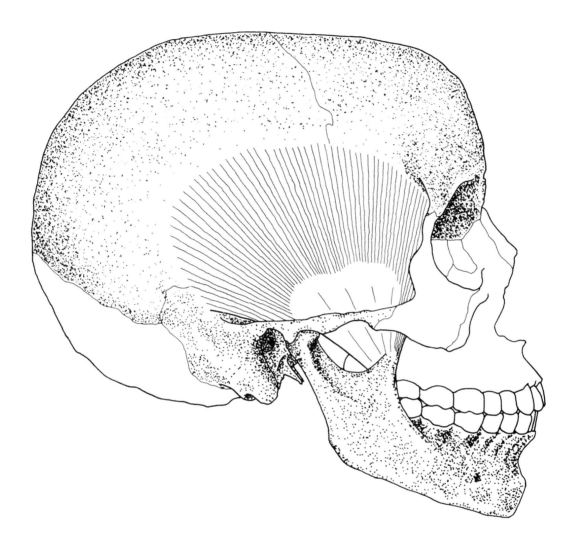

Skull—lateral view

■ Origin
Temporal fossa including frontal, parietal, and temporal bones

■ Insertion
Coronoid process and anterior border of ramus of mandible

■ Action
Closes lower jaw, clenches teeth

■ Nerve
Mandibular division of trigeminal nerve

MASSETER

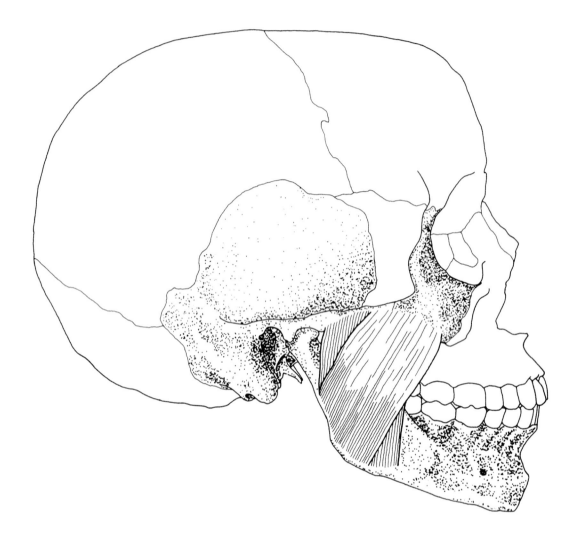

Skull—lateral view

■ Origin
Zygomatic process of maxilla, medial and inferior surfaces of zygomatic arch

■ Insertion
Angle and ramus of mandible, lateral surface of coronoid process of mandible

■ Action
Closes lower jaw, clenches teeth

■ Nerve
Mandibular division of trigeminal nerve

Note: Superficial fibers slightly protract jaw (see lateral pterygoid).

PTERYGOIDEUS MEDIALIS *(Medial Pterygoid)*

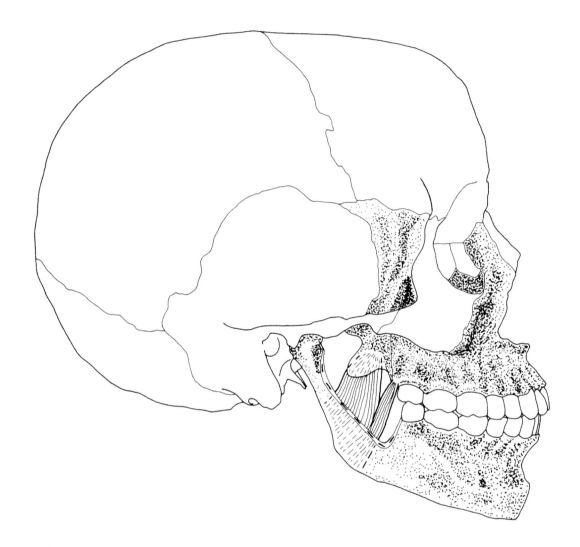

Skull—lateral view
(Part of mandible cut away)

■ Origin
Medial surface of lateral pterygoid plate of sphenoid bone, palatine bone, and tuberosity of maxilla

■ Insertion
Medial surface of ramus and angle of mandible

■ Action
Closes lower jaw, clenches teeth

■ Nerve
Mandibular division of trigeminal nerve

PTERYGOIDEUS LATERALIS   *(Lateral Pterygoid)*

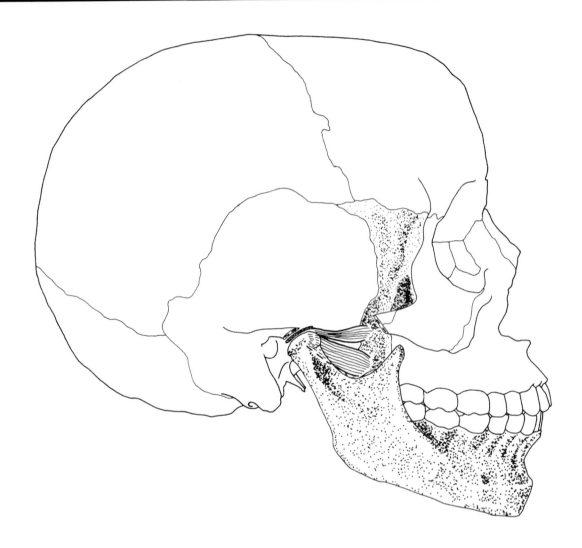

Skull—lateral view

■ Origin

Superior head*—lateral surface of greater wing of sphenoid

Inferior head—lateral surface of lateral pterygoid plate

■ Insertion

Condyle of mandible, temporomandibular joint

■ Action

Opens jaws, protrudes mandible, moves mandible sidewards

■ Nerve

Mandibular division of trigeminal nerve

Note: This sideward movement, aided by superficial fibers of masseter, causes chewing movements.

*Stern calls this a separate muscle: superior pterygoid.

Reference: Stern, J.T.: *Essentials of Gross Anatomy*, F.A. Davis Company, Philadelphia, 1988.

CHAPTER FOUR

Muscles of the Neck

STERNOCLEIDOMASTOID

Three-quarter frontal view

■ Origin
Sternal head—manubrium of sternum

Clavicular head—medial part of clavicle

■ Insertion
Mastoid process of temporal bone, lateral half of superior nuchal line of occipital bone

■ Action
One side—bends neck laterally, rotates head to opposite side

Both sides together—flexes neck, draws head ventrally and elevates chin, draws sternum superiorly in deep inspiration

■ Nerve
Spinal part of accessory nerve (C2, C3)

PLATYSMA

Three-quarter frontal view

■ Origin
Subcutaneous fascia of upper one-fourth of chest

■ Insertion
Subcutaneous fascia and muscles of chin and jaw, mandible

■ Action
Depresses and draws lower lip laterally, draws up skin of chest, depresses mandible

■ Nerve
Cervical branch of facial nerve

DIGASTRICUS

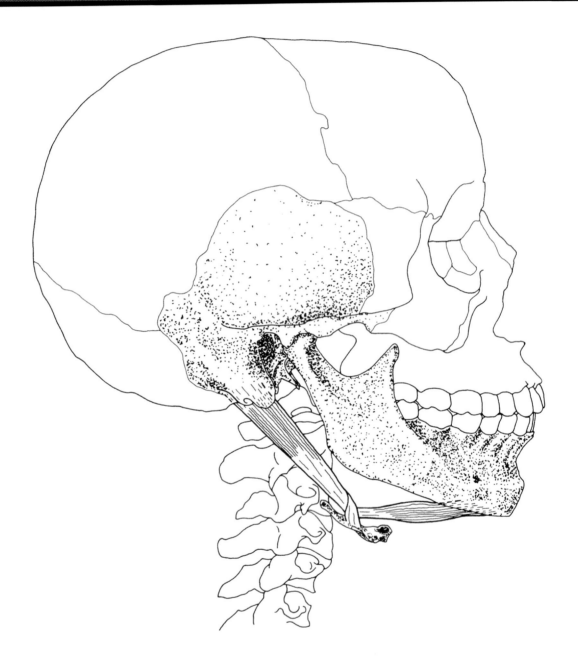

Lateral view

■ Origin
Posterior belly—mastoid notch of temporal bone

Anterior belly—inner side of inferior border of mandible near symphysis

■ Insertion
Intermediate tendon attached to hyoid bone

■ Action
Raises hyoid bone, depresses mandible, moves hyoid forward or backward

■ Nerve
Anterior belly—mandibular division of trigeminal

Posterior belly—facial nerve

STYLOHYOID

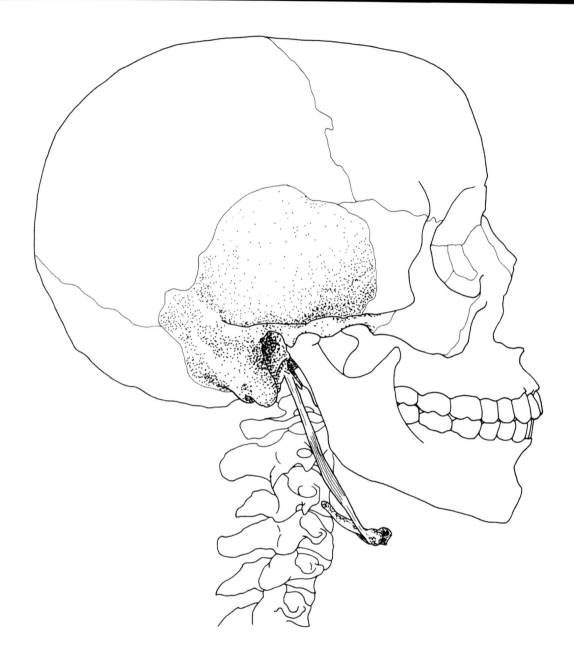

Lateral view

■ **Origin**
Styloid process of temporal bone

■ **Insertion**
Hyoid bone

■ **Action**
Draws hyoid bone backward, elevates tongue

■ **Nerve**
Facial nerve

MYLOHYOID

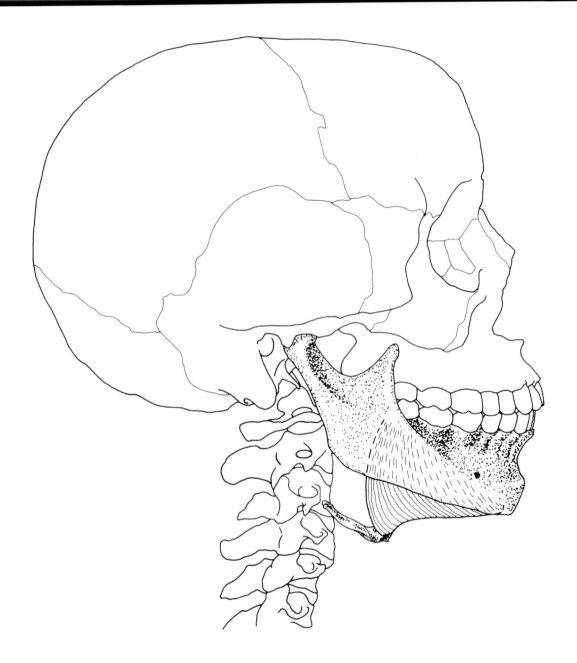

Lateral view

■ **Origin**
Inside surface of mandible from symphysis to molars
(mylohyoid line)

■ **Insertion**
Hyoid bone

■ **Action**
Elevates hyoid bone, raises floor of mouth and tongue

■ **Nerve**
Mandibular division of trigeminal nerve

GENIOHYOID

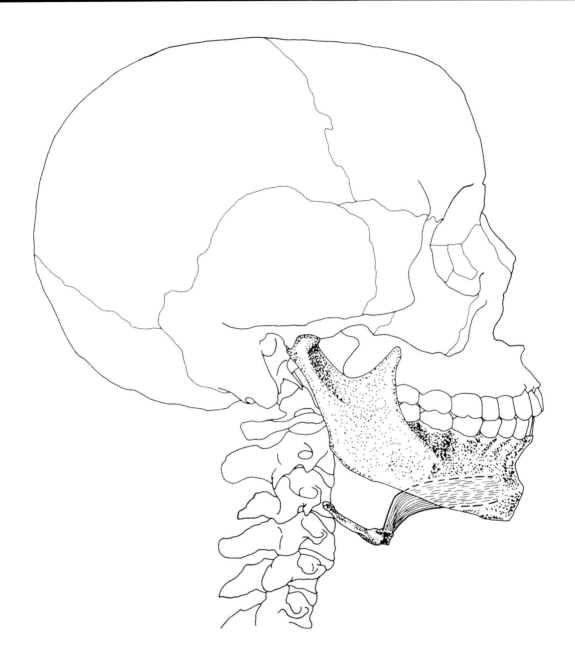

Lateral view

■ **Origin**
Inferior mental spine on interior medial surface of mandible

■ **Insertion**
Body of hyoid bone

■ **Action**
Protrudes hyoid bone and tongue

■ **Nerve**
Branch of C1 through hypoglossal nerve

STERNOHYOID

Frontal view

■ **Origin**
Medial end of clavicle, manubrium of sternum

■ **Insertion**
Body of hyoid bone

■ **Action**
Depresses hyoid bone

■ **Nerve**
Ansa cervicalis (C1–C3)

Lateral view

STERNOTHYROID

Lateral view

Frontal view

■ **Origin**
Dorsal surface of manubrium of sternum

■ **Insertion**
Lamina of thyroid cartilage

■ **Action**
Depresses thyroid cartilage

■ **Nerve**
Ansa cervicalis (C1–C3)

THYROHYOID

Lateral view

■ **Origin**
Lamina of thyroid cartilage

■ **Insertion**
Greater cornu of hyoid bone

■ **Action**
Depresses hyoid or raises thyroid

■ **Nerve**
C1 through hypoglossal nerve

OMOHYOID

Lateral view

■ **Origin**

Superior border of scapula

■ **Insertion**

Inferior belly—bound to clavicle by central tendon

Superior belly—continues to body of hyoid bone

■ **Action**

Depresses hyoid bone

■ **Nerve**

Ansa cervicalis (C2, C3)

LONGUS COLLI

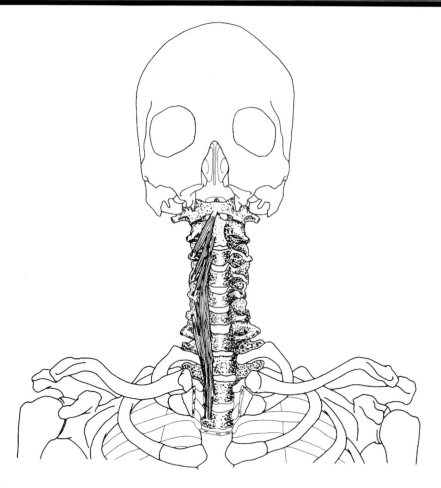

Frontal view
(Mandible and part of maxilla removed)

Superior oblique part

■ **Origin**
Transverse processes of third, fourth, and fifth cervical vertebrae

■ **Insertion**
Anterior arch of atlas

Inferior oblique part

■ **Origin**
Anterior surface of bodies of first two or three thoracic vertebrae

■ **Insertion**
Transverse processes of fifth and sixth cervical vertebrae

Vertical part

■ **Origin**
Anterior surfaces of bodies of upper three thoracic and lower three cervical vertebrae

■ **Insertion**
Anterior surfaces of the second, third, and fourth cervical vertebrae

■ **Action**
All three parts flex cervical vertebrae

■ **Nerve**
C2–C7

Note: Cervical hyperextension injuries (whiplash) may strain these muscles and/or sprain the anterior ligaments of vertebrae.

LONGUS CAPITIS

Frontal view
(Mandible and part of maxilla removed)

■ Origin
Anterior tubercles of transverse processes of third through sixth cervical vertebrae

■ Insertion
Occipital bone anterior to foramen magnum

■ Action
Acting together (bilaterally)—flex head

Acting on one side only—rotate head

■ Nerve
C1–C3

RECTUS CAPITIS ANTERIOR

Frontal view
(Mandible and part of maxilla removed)

■ **Origin**
Anterior base of transverse process of atlas

■ **Insertion**
Occipital bone anterior to foramen magnum

■ **Action**
Flexes head

■ **Nerve**
C2, C3

RECTUS CAPITIS LATERALIS

Frontal view
(Mandible and part of maxilla removed)

■ **Origin**
Transverse process of atlas

■ **Insertion**
Jugular process of occipital bone

■ **Action**
Bends head laterally

■ **Nerve**
C2, C3

SCALENUS ANTERIOR

Three-quarter frontal view
(Mandible and part of maxilla removed)

■ **Origin**
Transverse processes of third through sixth cervical vertebrae

■ **Insertion**
Inner border of first rib (scalene tubercle)

■ **Action**
Raises first rib (respiratory inspiration); acting together, they flex neck; acting on one side, they laterally flex, rotate neck

■ **Nerve**
Ventral rami of cervical nerves

SCALENUS MEDIUS

Three-quarter frontal view
(Mandible and part of maxilla removed)

■ Origin
Transverse processes of lower six cervical vertebrae
(C2–C7)

■ Insertion
Upper surface of first rib

■ Action
Raises first rib (respiratory inspiration); acting together,
they flex neck; acting on one side, they laterally flex,
rotate neck

■ Nerve
Ventral rami of cervical nerves

SCALENUS POSTERIOR

Three-quarter frontal view
(Mandible and part of maxilla removed)

■ **Origin**
Transverse processes of lower two or three cervical vertebrae (C5–C7)

■ **Insertion**
Outer surface of second rib

■ **Action**
Raises second rib (respiratory inspiration); acting together, they flex neck; acting on one side, they laterally flex, rotate neck

■ **Nerve**
Ventral rami of lower cervical nerves

RECTUS CAPITIS POSTERIOR MAJOR

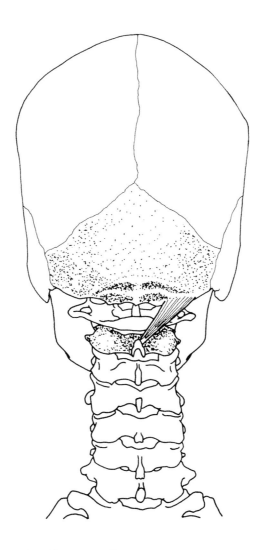

Posterior skull and cervical vertebrae

■ Origin
Spinous process of axis

■ Insertion
Lateral portion of inferior nuchal line of occipital bone

■ Action
Extends and rotates head

■ Nerve
Suboccipital nerve

RECTUS CAPITIS POSTERIOR MINOR

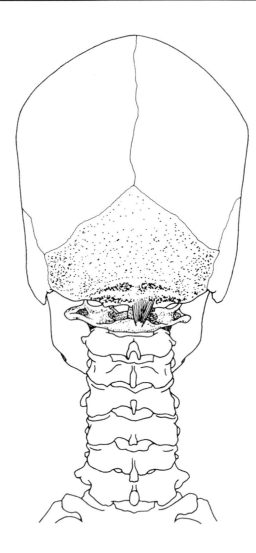

Posterior skull and cervical vertebrae

■ Origin
Posterior arch of atlas

■ Insertion
Medial portion of inferior nuchal line of occipital bone

■ Action
Extends head

■ Nerve
Suboccipital nerve

OBLIQUUS CAPITIS INFERIOR

Posterior skull and cervical vertebrae

■ **Origin**
Spinous process of axis

■ **Insertion**
Transverse process of atlas

■ **Action**
Rotates atlas

■ **Nerve**
Suboccipital nerve

OBLIQUUS CAPITIS SUPERIOR

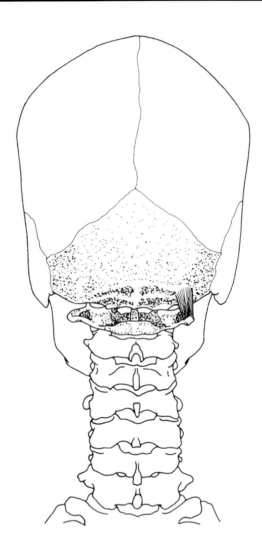

Posterior skull and cervical vertebrae

■ Origin
Transverse process of atlas

■ Insertion
Occipital bone between inferior and superior nuchal lines

■ Action
Extends and bends head laterally

■ Nerve
Suboccipital nerve

CHAPTER FIVE

Muscles of
the Trunk

SPLENIUS CAPITIS

Posterior skull, neck, and back

■ Origin
Lower part of ligamentum nuchae, spinous processes of seventh cervical vertebra (C7) and upper three or four thoracic vertebrae (T1–T4)

■ Insertion
Mastoid process of temporal bone and lateral part of superior nuchal line

■ Action
Acting together, they extend, hyperextend head, neck; acting on one side, they laterally flex, rotate head, neck

■ Nerve
Lateral branches of dorsal primary divisions of middle and lower cervical nerves

SPLENIUS CERVICIS

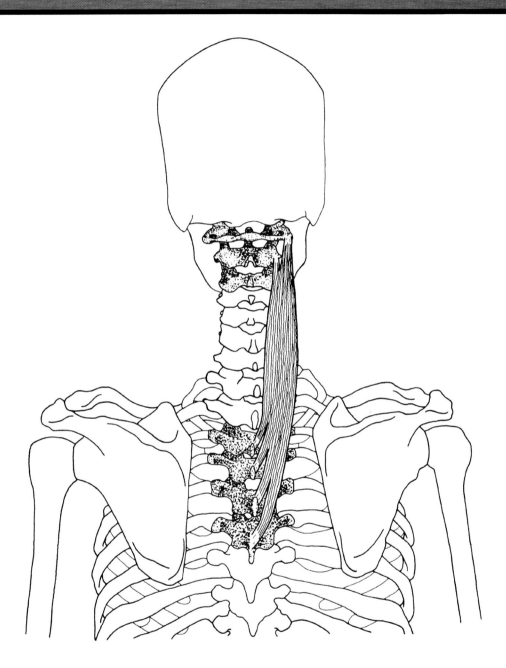

Posterior skull, neck, and back

■ **Origin**
Spinous processes of third through sixth thoracic vertebrae (T3–T6)

■ **Insertion**
Transverse processes of upper two or three cervical vertebrae (C1–C3)

■ **Action**
Acting together, they extend, hyperextend head, neck; acting on one side, they laterally flex, rotate head, neck

■ **Nerve**
Lateral branches of dorsal primary divisions of middle and lower cervical nerves

ERECTOR SPINAE

Iliocostalis cervicis

Origin
Angles of third through sixth ribs

Insertion
Transverse processes of fourth, fifth, and sixth cervical vertebrae

Action
Extension, lateral flexion of vertebral column

Nerve
Dorsal primary divisions of spinal nerves

Iliocostalis thoracis

Origin
Angles of lower six ribs medial to iliocostalis lumborum

Insertion
Angles of upper six ribs and transverse process of seventh cervical vertebra

Action
Extension, lateral flexion of vertebral column, rotates ribs for forceful inspiration

Nerve
Dorsal primary divisions of spinal nerves

Iliocostalis lumborum

Origin
Medial and lateral sacral crests and medial part of iliac crests

Insertion
Angles of lower six ribs

Action
Extension, lateral flexion of vertebral column, rotates ribs for forceful inspiration

Nerve
Dorsal primary divisions of spinal nerves

*The erector spinae (sacrospinalis) is a complex of three sets of muscles: iliocostalis, longissimus, and spinalis. The origin of this group is the medial and lateral sacral crests, the medial part of iliac crests, and the spinous processes and supraspinal ligament of lumbar and eleventh and twelfth thoracic vertebrae.

Trunk—dorsal view

ERECTOR SPINAE

Trunk—dorsal view

Longissimus capitis

■ **Origin**
Transverse processes of upper five thoracic vertebrae (T1–T5), articular processes of lower three cervical vertebrae (C5–C7)

■ **Insertion**
Posterior part of mastoid process of temporal bone

■ **Action**
Extends and rotates head

■ **Nerve**
Dorsal primary divisions of middle and lower cervical nerves

Longissimus cervicis

■ **Origin**
Transverse processes of upper four or five thoracic vertebrae (T1–T5)

■ **Insertion**
Transverse processes of second through sixth cervical vertebrae

■ **Action**
Extension, lateral flexion of vertebral column

■ **Nerve**
Dorsal primary divisions of spinal nerves

Longissimus thoracis

■ **Origin**
Medial and lateral sacral crests, spinous processes and supraspinal ligament of lumbar and eleventh and twelfth thoracic vertebrae, and medial part of iliac crests

■ **Insertion**
Transverse processes of all thoracic vertebrae, between tubercles and angles of lower nine or ten ribs

■ **Action**
Extension, lateral flexion of vertebral column, rotates ribs for forceful inspiration

■ **Nerve**
Dorsal primary divisions of spinal nerves

ERECTOR SPINAE

Spinalis capitis
(Medial part of semispinalis capitis)

Spinalis cervicis

Origin
Ligamentum nuchae, spinous process of seventh cervical vertebra

Insertion
Spinous process of axis

Action
Extends vertebral column

Nerve
Dorsal primary divisions of spinal nerves

Spinalis thoracis

Origin
Spinous processes of lower two thoracic (T11, T12) and upper two lumbar (L1, L2) vertebrae

Insertion
Spinous processes of upper thoracic vertebrae (T1–T8)

Action
Extends vertebral column

Nerve
Dorsal primary divisions of spinal nerves

Trunk—dorsal view

TRANSVERSOSPINALIS*

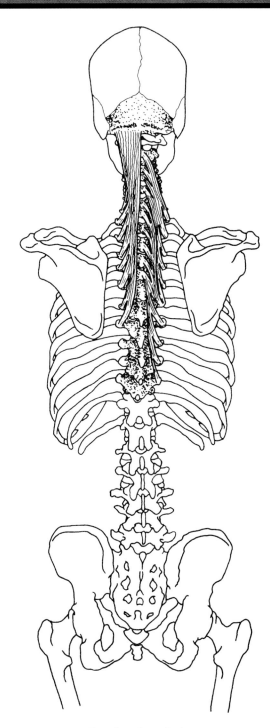

Trunk—dorsal view

Semispinalis capitis
(Medial part is spinalis capitis)

■ **Origin**
Transverse processes of lower four cervical (C4–C7) and upper six or seven thoracic (T1–T7) vertebrae

■ **Insertion**
Between superior and inferior nuchal lines of occipital bone

■ **Action**
Extends and rotates head

■ **Nerve**
Dorsal primary divisions of spinal nerves

Semispinalis cervicis

■ **Origin**
Transverse processes of upper five or six thoracic vertebrae (T1–T6)

■ **Insertion**
Spinous processes of second to fifth cervical vertebrae (C2–C5)

■ **Action**
Extends and rotates vertebral column

■ **Nerve**
Dorsal primary divisions of spinal nerves

Semispinalis thoracis

■ **Origin**
Transverse processes of the sixth through tenth thoracic vertebrae (T6–T10)

■ **Insertion**
Spinous processes of the lower two cervical (C6, C7) and upper four thoracic (T1–T4) vertebrae

■ **Action**
Extends and rotates vertebral column

■ **Nerve**
Dorsal primary divisions of spinal nerves

*The transversospinalis is composed of groups of small muscles generally extending upward from transverse processes to spinous processes of higher vertebrae. They are deep to erector spinae. They include semispinalis, multifidi, and rotatores.

MULTIFIDIS*

Trunk—dorsal view

■ Origin

Sacral region—along sacral foramina up to posterior superior iliac spine

Lumbar region—mammillary processes[†] of vertebrae

Thoracic region—transverse processes

Cervical region—articular processes of lower four vertebrae (C4–C7)

■ Insertion

Spinous process two to four vertebrae superior to origin

■ Action

Extend and rotate vertebral column

■ Nerve

Dorsal primary division of spinal nerves

*Part of transversospinalis.
†Posterior border of superior articular process.

ROTATORES*

Trunk—dorsal view

■ Origin
Transverse process of each vertebra

■ Insertion
Base of spinous process of next vertebra above

■ Action
Extend and rotate vertebral column

■ Nerve
Dorsal primary division of spinal nerves

*Part of transversospinalis.

INTERSPINALES *(Paired on either side of interspinal ligament)*

Trunk—dorsal view

■ Origin
Cervical region—spinous processes of third to seventh cervical vertebrae (C3–C7)

Thoracic region—spinous processes of second to twelfth thoracic vertebrae (T2–T12)

Lumbar region—spinous processes of second to fifth lumbar vertebrae (L2–L5)

■ Insertion
Spinous process of next vertebra superior to origin

■ Action
Extend vertebral column

■ Nerve
Dorsal primary division of spinal nerves

INTERTRANSVERSARII

Trunk—dorsal view

Cervical region

Intertransversarii anteriores

- **Origin**   Anterior tubercle of transverse processes of vertebrae from first thoracic to axis
- **Insertion**   Anterior tubercle of next superior vertebra
- **Action**   Lateral flexion of vertebral column
- **Nerve**   Ventral primary division of spinal nerves

Intertransversarii posteriores

- **Origin**   Posterior tubercle of transverse processes of vertebrae from first thoracic to axis
- **Insertion**   Posterior tubercle of next superior vertebra

Thoracic region

- **Origin**   Transverse processes of first lumbar to eleventh thoracic vertebrae
- **Insertion**   Transverse processes of next superior vertebra

Lumbar region

Intertransversarii laterales

- **Origin**   Transverse processes of lumbar vertebrae
- **Insertion**   Transverse process of next superior vertebra
- **Action**   Lateral flexion of vertebral column
- **Nerve**   Ventral primary division of spinal nerves

Intertransversarii mediales

- **Origin**   Mammillary process* of each lumbar vertebra
- **Insertion**   Accessory process of the next superior lumbar vertebra
- **Action**   Lateral flexion of vertebral column
- **Nerve**   Dorsal primary division of spinal nerves

*Posterior border of superior articular process.

INTERCOSTALES EXTERNI *(External Intercostal)*

Trunk—anterior view

■ Origin
Lower margin of upper eleven ribs

■ Insertion
Superior border of rib below (each muscle fiber runs obliquely and inserts toward the costal cartilage)

■ Action
Draw ventral part of ribs upward, increasing the volume of the thoracic cavity for inspiration

■ Nerve
Intercostal nerves

INTERCOSTALES INTERNI *(Internal Intercostal)*

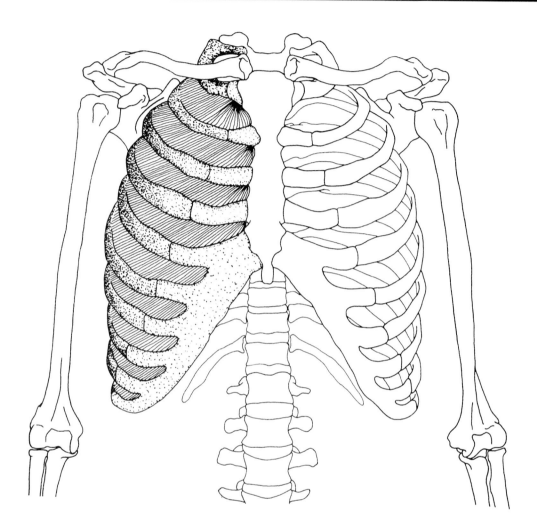

Trunk—anterior view

■ Origin
From the cartilages to the angles of the upper eleven ribs

■ Insertion
Superior border of the rib below (each muscle fiber runs obliquely and inserts away from the costal cartilage)

■ Action
Draw ventral part of ribs downward, decreasing the volume of the thoracic cavity for expiration

■ Nerve
Intercostal nerves

SUBCOSTALES

Trunk—dorsal view

■ Origin
Inner surface of each rib near its angle

■ Insertion
Medially on the inner surface of second or third rib below

■ Action
Draw ventral part of ribs downward, decreasing the volume of the thoracic cavity for forceful expiration

■ Nerve
Intercostal nerves

Note: These muscles are deep to the internal intercostals. They continue distally between single ribs, where they are known as innermost intercostal muscles.

TRANSVERSUS THORACIS

Trunk—anterior view

■ Origin
Inner surface of lower portion of sternum and adjacent costal cartilages

■ Insertion
Inner surfaces of costal cartilages of the second through sixth ribs

■ Action
Draws ventral part of ribs downward, decreasing the volume of the thoracic cavity for forceful expiration

■ Nerve
Intercostal nerves

Note: These muscles are deep to the internal intercostal muscles.

LEVATORES COSTARUM

Trunk—dorsal view

■ Origin
Transverse processes of the seventh cervical and the upper eleven thoracic vertebrae

■ Insertion
Laterally to outer surface of next lower rib (lower muscles may cross over one rib)

■ Action
Raises ribs; extends, laterally flexes, and rotates vertebral column

■ Nerve
Intercostal nerves

SERRATUS POSTERIOR SUPERIOR

Trunk—dorsal view

■ **Origin**

Ligamentum nuchae, spinous processes of seventh cervical and first few thoracic vertebrae

■ **Insertion**

Upper borders of the second through fifth ribs lateral to their angles

■ **Action**

Raises ribs in inspiration

■ **Nerve**

T1–T4

SERRATUS POSTERIOR INFERIOR

Trunk—dorsal view

■ **Origin**
Spinous processes of the lower two thoracic and the upper two or three lumbar vertebrae

■ **Insertion**
Lower borders of bottom four ribs

■ **Action**
Pulls ribs down, resisting pull of diaphragm

■ **Nerve**
T9–T12

DIAPHRAGM

Trunk—anterior view
(Lower costal cartilages removed)

■ Origin
Sternal part—inner part of xiphoid process

Costal part—inner surfaces of lower six ribs and their cartilages

Lumbar part—upper two or three lumbar vertebrae and lateral and medial lumbocostal arches*

■ Insertion
Fibers converge and meet on a central tendon

■ Action
Draws central tendon inferiorly, for inspiration

■ Nerve
Phrenic nerve (C3–C5)

Note: This muscle inserts upon itself. Its action is to change the volume of the thoracic and abdominal cavities.

*These tendinous structures, also known as the medial and lateral arcuate ligaments, allow the diaphragm to bridge the upper parts of the psoas major and quadratus lumborum muscles.

OBLIQUUS EXTERNUS ABDOMINIS *(External Oblique)*

Trunk—lateral view

■ Origin
Lower eight ribs

■ Insertion
Anterior part of iliac crest, abdominal aponeurosis to linea alba

■ Action
Compresses abdominal contents, laterally flexes and rotates vertebral column

■ Nerve
Eighth to twelfth intercostal, iliohypogastric, ilioinguinal nerves

■ Relationships
Most superficial of the three lateral abdominal muscles

Note: Important in forced expiration, coughing, sneezing.

OBLIQUUS INTERNUS ABDOMINIS *(Internal Oblique)*

Trunk—lateral view

■ Origin
Lateral half of inguinal ligament, iliac crest, thoracolumbar fascia

■ Insertion
Cartilage of bottom three or four ribs, abdominal aponeurosis to linea alba

■ Action
Compresses abdominal contents, laterally flexes and rotates vertebral column

■ Nerve
Eighth to twelfth intercostal, iliohypogastric, ilioinguinal nerves

■ Relationships
Middle layer of the three lateral abdominal muscles

Note: Important in forced expiration, coughing, sneezing.

CREMASTER

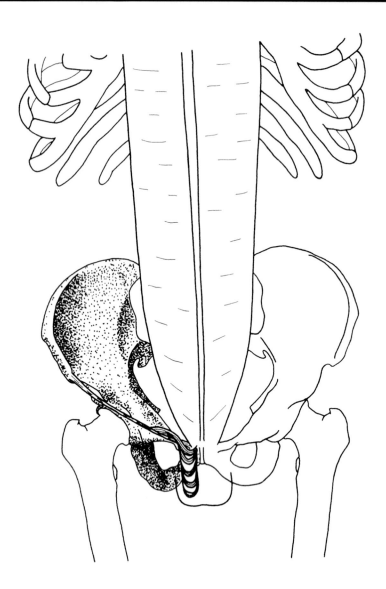

Trunk—anterior view

■ **Origin**
Inguinal ligament

■ **Insertion**
Pubic tubercle, crest of pubis, sheath of rectus abdominis

■ **Action**
Pulls testes toward body

■ **Nerve**
Genital branch of genitofemoral nerve

Note: The cremaster regulates the temperature of the testes, which is important for spermatogenesis.

TRANSVERSUS ABDOMINIS

Trunk—lateral view

■ **Origin**
Lateral part of inguinal ligament, iliac crest, thoracolumbar fascia, cartilage of lower six ribs

■ **Insertion**
Abdominal aponeurosis to linea alba

■ **Action**
Compresses abdomen

■ **Nerve**
Seventh to twelfth intercostal, iliohypogastric, ilioinguinal nerves

■ **Relationships**
Deepest of the three lateral abdominal muscles

Note: Important in forced expiration, coughing, sneezing.

RECTUS ABDOMINIS*

Trunk—anterior view

■ Origin
Crest of pubis, pubic symphysis

■ Insertion
Cartilage of fifth, sixth, and seventh ribs, xiphoid process

■ Action
Flexes vertebral column, compresses abdomen

■ Nerve
Seventh through twelfth intercostal nerves

*Tendinous bands divide each rectus into three or four bellies. Each rectus is sheathed in aponeurotic fibers from the lateral abdominal muscles. These fibers meet centrally to form the linea alba.

Note: The pyramidalis is a small, unimportant muscle that extends from the ventral surface of the pubis to the lower part of the linea alba. It is frequently absent.

ABDOMINAL MUSCLES

Trunk—anterior and cross-sectional views

1. Obliquus externus abdominis
2. Obliquus internus abdominis
3. Transversus abdominis
4. Rectus abdominis
5. Linea alba

Note: The aponeuroses (broad, flat tendons) of the three lateral abdominal muscles join to form the fascial sheath surrounding the rectus abdominis.

QUADRATUS LUMBORUM

Lower trunk—anterior view

■ Origin
Iliolumbar ligament, iliac crest

■ Insertion
Twelfth rib, transverse processes of upper four lumbar vertebrae

■ Action
Laterally flexes vertebral column, fixes ribs for forced expiration

■ Nerve
T12, L1

Note: Fixation of the ribs may provide a stable attachment of the diaphragm for voice control in singers.

CHAPTER SIX

Muscles of the Shoulder and Arm

PECTORALIS MAJOR

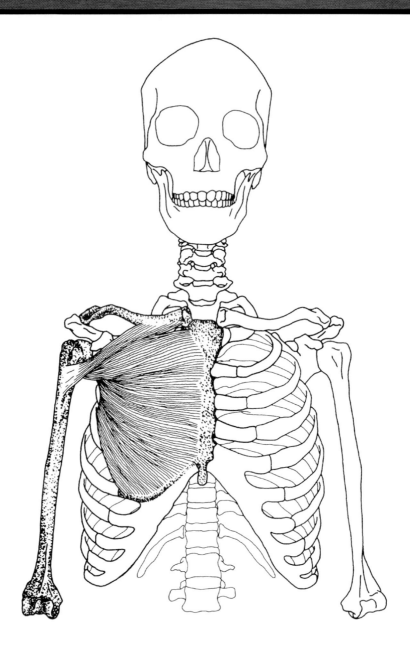

Anterior view

■ Origin

Clavicular part—medial half of the clavicle

Sternocostal part—sternum, upper six costal cartilages, aponeurosis of external oblique

■ Insertion

Lateral lip of intertubercular (bicipital) groove of humerus, crest below greater tubercle of the humerus

■ Action

Both parts adduct, medially rotate arm; clavicular part flexes arm from full extension; sternocostal part extends the flexed arm

■ Nerve

Medial and lateral pectoral nerves (C5–C8, T1)

PECTORALIS MINOR

Anterior view

■ **Origin**
External surfaces of the third, fourth, and fifth ribs

■ **Insertion**
Coracoid process of the scapula

■ **Action**
Draws scapula forward and downward, raises ribs* in forced inspiration

■ **Nerve**
Medial pectoral nerve (C8, T1)

■ **Relationships**
Deep to pectoralis major

*Raising the ribs requires stabilization of the scapula by the rhomboids and trapezius.

SUBCLAVIUS

Anterior view

■ Origin
Junction of the first rib with its costal cartilage

■ Insertion
Groove on the inferior (lower) surface of the clavicle

■ Action
Depresses clavicle, draws shoulder forward and downward, steadies clavicle during movements of shoulder girdle

■ Nerve
C5, C6

CORACOBRACHIALIS

Anterior view

■ **Origin**
Tip (apex) of the coracoid process of scapula

■ **Insertion**
Middle third of the medial surface and border of the humerus

■ **Action**
Weakly adducts arm (flexion unsubstantiated), aids in stabilizing humerus

■ **Nerve**
Musculocutaneous nerve (C6, C7)

■ **Relationships**
Deep to short head of biceps

BICEPS BRACHII

Anterior view

■ Origin
Long head—supraglenoid tubercle of scapula

Short head—coracoid process of scapula

■ Insertion
Tuberosity of radius, bicipital aponeurosis into deep fascia on medial part of forearm

■ Action
Supinates forearm, flexes forearm, weakly flexes arm at shoulder

■ Nerve
Musculocutaneous nerve (C5, C6)

■ Relationships
Long head passes through intertubercular (bicipital) groove, then inside glenohumeral joint capsule

BRACHIALIS

Anterior view

■ **Origin**
Anterior of lower half of humerus

■ **Insertion**
Coronoid process of ulna, tuberosity of ulna

■ **Action**
Flexes forearm

■ **Nerve**
Musculocutaneous nerve (C5, C6)

■ **Relationships**
Deep to biceps brachii

MUSCLES OF THE ANTERIOR CHEST AND ARM

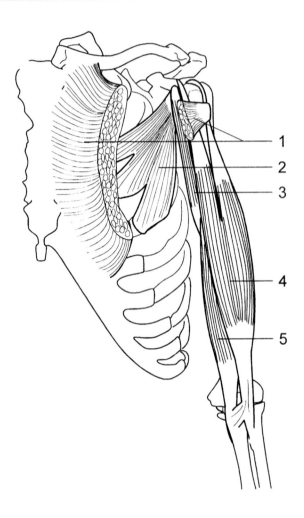

Shoulder—anterior view

1. Pectoralis major (cut)
2. Pectoralis minor
3. Coracobrachialis
4. Biceps brachii
5. Brachialis

TRAPEZIUS

Posterior view

■ Origin
Medial third of superior nuchal line, external occipital protuberance, ligamentum nuchae, spinous processes and supraspinous ligaments of seventh cervical and all thoracic vertebrae

■ Insertion
Upper part—lateral third of clavicle

Middle part—acromion and crest of spine of scapula

Lower part—medial portion of crest of spine of scapula (tubercle)

■ Action
Upper part elevates scapula,* middle part retracts (adducts) scapula, lower part depresses scapula, upper and lower parts together rotate scapula (important in elevating arm)

■ Nerve
Accessory (eleventh cranial), C3, C4

■ Relationships
Most superficial muscle of back

*Upper part stabilizes scapula against downward rotation, as when weight is carried in the hand.

LATISSIMUS DORSI

Posterior view

■ Origin
Spinous processes of the lower six thoracic vertebrae, lumbar vertebrae, sacral vertebrae, supraspinal ligament, and posterior part of the iliac crest through the lumbar (thoracolumbar) fascia, lower three or four ribs, inferior angle of the scapula

■ Insertion
Bottom of intertubercular (bicipital) groove

■ Action
Extends, adducts, and medially rotates the arm, draws the shoulder downward and backward, keeps inferior angle of scapula against the chest wall, accessory muscle of respiration

■ Nerve
Thoracodorsal nerve (C6–C8)

Note: This muscle is used for the crawl stroke in swimming.

LEVATOR SCAPULAE

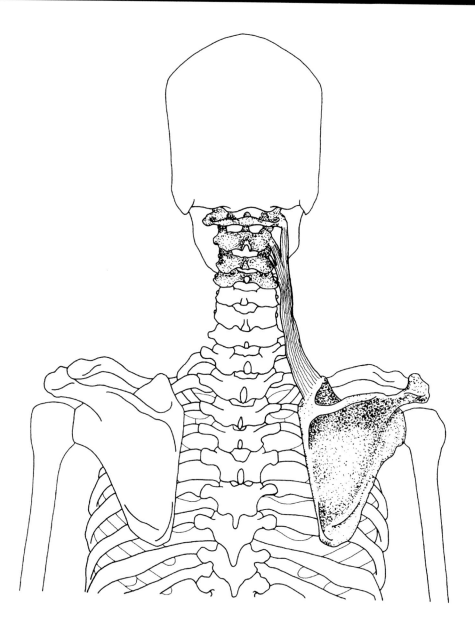

Posterior view

■ Origin
Posterior tubercles of the transverse processes of the first four cervical vertebrae

■ Insertion
Vertebral (medial) border of the scapula at and above the spine

■ Action
Elevates medial border of scapula, rotates scapula to lower the lateral angle, acts with trapezius and rhomboids to pull scapula medially and upward, bends neck laterally

■ Nerve
Dorsal scapular nerve (C5)

RHOMBOID MAJOR

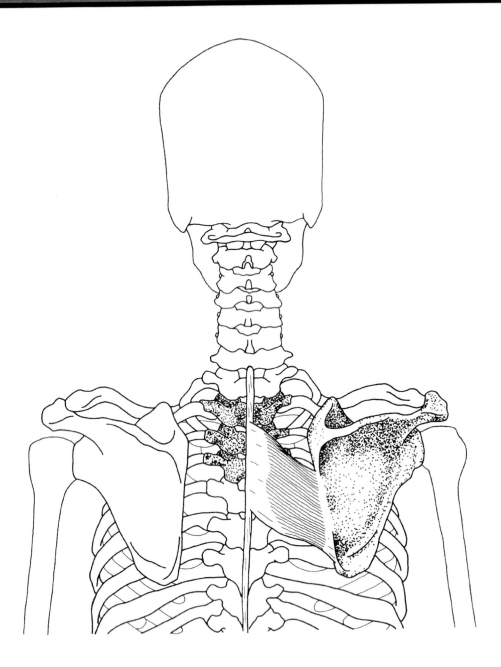

Posterior view

■ Origin
Spines of the second to fifth thoracic vertebrae, supraspinous ligament

■ Insertion
Medial border of the scapula below the spine

■ Action
Retracts and stabilizes scapula, elevates the medial border of the scapula causing downward rotation, assists in adduction of arm

■ Nerve
Dorsal scapular nerve (C5)

RHOMBOID MINOR

Posterior view

■ Origin
Spines of the seventh cervical and first thoracic
vertebrae, lower part of the ligamentum nuchae

■ Insertion
Medial border of the scapula at the root of the spine

■ Action
Retracts and stabilizes scapula, elevates the medial
border of the scapula, rotates the scapula to depress the
lateral angle (assists in adduction of arm)

■ Nerve
Dorsal scapular nerve (C5)

Serratus Anterior

■ Origin
Outer surfaces and superior borders of first eight or nine ribs, and fascia covering first intercostal space

■ Insertion
Anterior surface (costal surface) of the medial border of the scapula

■ Action
Rotates scapula for abduction and flexion of arm, protracts scapula

■ Nerve
Long thoracic nerve (C5–C7)

■ Relationships
Serratus anterior and rhomboids both insert on the medial border of scapula; they are antagonists causing protraction and retraction

Lateral view

DELTOID

Lateral view

■ Origin
Anterior portion—anterior border and superior surface of the lateral third of the clavicle

Middle portion—lateral border of the acromion process

Posterior portion—lower border of the crest of the spine of the scapula

■ Insertion
Deltoid tuberosity, on the middle of the lateral surface of the shaft of the humerus

■ Action
Anterior portion—flexes and medially rotates arm

Middle portion—abducts arm

Posterior portion—extends and laterally rotates arm

■ Nerve
Axillary nerve (C5, C6)

SUPRASPINATUS *(Rotator cuff*)*

Lateral view

■ Origin
Supraspinous fossa of scapula

■ Insertion
Upper part of the greater tuberosity of the humerus, capsule of the shoulder joint

■ Action
Aids deltoid in abduction of arm; draws humerus toward glenoid fossa, preventing deltoid from forcing humerus up against acromion; weakly flexes arm

■ Nerve
Suprascapular nerve (C5)

*Supraspinatus, infraspinatus, teres minor, and subscapularis together are called the rotator cuff. They prevent the larger muscles from dislocating the humerus during their actions.

Note: Because of the great range of motion between the humerus and scapula, the joint capsule and its ligaments cannot provide necessary support. The rotator cuff muscles prevent dislocation of the humerus throughout most of the arm's range of motion.

INFRASPINATUS *(Rotator cuff)*

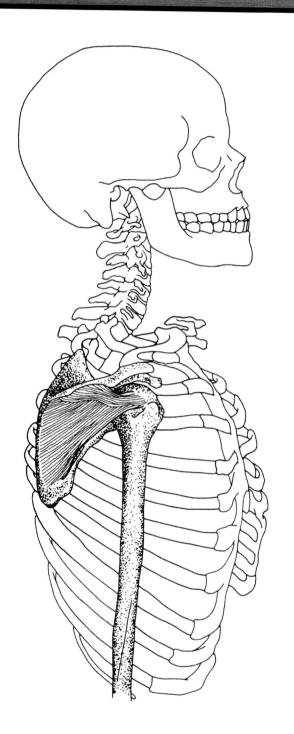

Lateral view

■ Origin
Infraspinous fossa of the scapula

■ Insertion
Middle facet of the greater tuberosity of the humerus, capsule of the shoulder joint

■ Action
Draws humerus toward glenoid fossa, thus resisting posterior dislocation of arm, as in crawling; laterally rotates; abducts arm

■ Nerve
Suprascapular nerve (C5, C6)

TERES MINOR *(Rotator cuff)*

Lateral view

■ Origin
Upper two-thirds of the dorsal surface of the axillary border of the scapula

■ Insertion
The capsule of the shoulder joint, the lower facet of the greater tuberosity of the humerus

■ Action
Laterally rotates arm, weakly adducts arm, draws humerus toward glenoid fossa

■ Nerve
Axillary nerve (C5)

SUBSCAPULARIS *(Rotator Cuff)*

Anterior view
(Upper ribs cut away)

■ Origin
Subscapular fossa on the anterior surface of scapula

■ Insertion
Lesser tuberosity of the humerus, ventral part of the capsule of the shoulder joint

■ Action
Medially rotates arm, stabilizes glenohumeral joint

■ Nerve
Upper and lower subscapular nerves (C5, C6)

TERES MAJOR

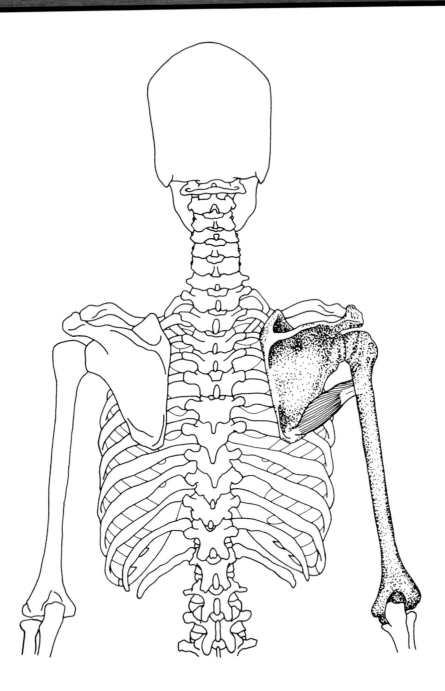

Posterior view

■ **Origin**
Lower third of the posterior surface of the lateral border of the scapula, near the inferior angle

■ **Insertion**
Medial lip of the bicipital groove of the humerus

■ **Action**
Medially rotates arm, adducts arm, extends arm

■ **Nerve**
Lower subscapular nerve (C5, C6)

TRICEPS BRACHII

Posterior view

■ Origin
Long head—infraglenoid tubercle of the scapula

Lateral head—upper half of the posterior surface of the shaft of the humerus

Medial head—posterior surface of the lower half of the shaft of the humerus

■ Insertion
Posterior part of olecranon process of the ulna

■ Action
Extends forearm, long head aids in adduction if arm is abducted

■ Nerve
Radial nerve (C7, C8)

Note: The radial nerve comes from the axilla (armpit) and passes along the humerus between the medial and lateral heads. Because it can be compressed against the humerus, it is the most commonly injured peripheral nerve.

ANCONEUS

Posterior view of arm

■ **Origin**
Posterior part of lateral epicondyle of the humerus

■ **Insertion**
Lateral surface of the olecranon process and posterior surface of ulna

■ **Action**
Extends forearm (assists triceps)

■ **Nerve**
Radial nerve (C7, C8)

POSTERIOR BACK, SHOULDER, AND ARM MUSCLES

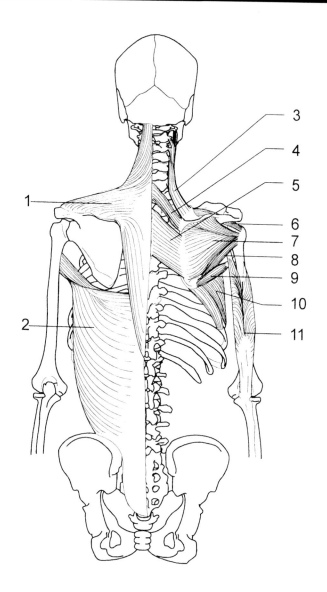

Trunk—posterior view

Superficial layer

1. Trapezius
2. Latissimus dorsi

Deep layer

3. Levator scapulae
4. Rhomboid minor
5. Rhomboid major

6. Supraspinatus (rotator cuff)
7. Infraspinatus (rotator cuff)
8. Teres minor (rotator cuff)
9. Teres major
10. Serratus anterior

Posterior arm

11. Triceps brachii

CHAPTER SEVEN

Muscles of the Forearm and Hand

PRONATOR TERES *(Superficial group)*

Forearm—anterior view

■ Origin
Humeral head—medial supracondylar ridge and medial epicondyle of the humerus

Ulnar head—medial border of the coronoid process of the ulna

■ Insertion
Middle of lateral surface of the radius (pronator tuberosity)

■ Action
Pronates and flexes forearm

■ Nerve
Median nerve (C6, C7)

FLEXOR CARPI RADIALIS *(Superficial group)*

Forearm—anterior view

■ Origin
Medial epicondyle of the humerus through the common tendon

■ Insertion
Front of the bases of the second and third metacarpal bones

■ Action
Flexes hand, synergist in abduction with extensor carpi radialis longus and brevis

■ Nerve
Median nerve (C6, C7)

PALMARIS LONGUS *(Superficial group)*

Forearm—anterior view

■ Origin
Medial epicondyle of the humerus through the common tendon

■ Insertion
Front (central part) of the flexor retinaculum and apex of the palmar aponeurosis

■ Action
Flexes the hand

■ Nerve
Median nerve (C6, C7)

Note: This muscle is absent in about 14% of limbs.

Reference: Agur, Amr: *Grant's Atlas of Anatomy,* 9th ed. Williams & Wilkins, Baltimore, 1991.

FLEXOR CARPI ULNARIS *(Superficial group)*

Forearm—anterior view

■ Origin
Humeral head—medial epicondyle of the humerus through the common tendon

Ulnar head—medial margin of olecranon process of ulna, dorsal border of shaft of the ulna

■ Insertion
Pisiform bone, hook of the hamate, and base of the fifth metacarpal bone

■ Action
Flexes hand, synergist in adduction of hand with extensor carpi ulnaris

■ Nerve
Ulnar nerve (C8, T1)

MUSCLES OF THE WRIST

Forearm—anterior view

1. Pronator teres
2. Flexor carpi radialis
3. Palmaris longus
4. Flexor carpi ulnaris
5. Flexor retinaculum
6. Palmar aponeurosis

FLEXOR DIGITORUM SUPERFICIALIS

Forearm—anterior view

■ Origin
Humeroulnar head—medial epicondyle of the humerus through common tendon,* medial margin of the coronoid process of ulna

Radial head—anterior surface of shaft of radius

■ Insertion
Four tendons divide into two slips each; slips insert into the sides (margins of the anterior surfaces) of the middle phalanges of four fingers

■ Action
Flexes the middle phalanges of the fingers

■ Nerve
Median nerve (C7, C8, T1)

■ Relationships
Deep to superficial flexors

*See superficial flexors.

Note: The tendons of flexor digitorum superficialis split and attach to the middle phalanx. The tendons of flexor digitorum profundus pass through this split and continue to the distal phalanx.

FLEXOR DIGITORUM PROFUNDUS

Forearm—anterior view

■ Origin
Upper three-fourths of anterior and medial surfaces of shaft of ulna and medial side of the coronoid process, interosseous membrane

■ Insertion
Front of base of distal phalanges of fingers

■ Action
Flexes distal phalanges

■ Nerve
Ulnar nerve supplies the medial half of the muscle (going to the little and ring fingers) (C8, T1)

Anterior interosseous branch of median nerve supplies lateral half (going to index and middle fingers) (C8, T1)

■ Relationships
Deep to flexor digitorum superficialis

Note: Both flexor digitorum muscles and the median nerve pass under the flexor retinaculum (p. 126) in the wrist. When irritated, the synovial sheaths of these muscles can compress the median nerve, causing the sensory and motor deficits known as carpal tunnel syndrome.

FLEXOR POLLICIS LONGUS

Forearm—anterior view

■ **Origin**
Middle of anterior surface of shaft of radius, interosseous membrane, medial epicondyle of humerus, and often coronoid process of ulna

■ **Insertion**
Palmar aspect of base of the distal phalanx of thumb

■ **Action**
Flexes the thumb

■ **Nerve**
Anterior interosseous branch of median nerve (C8, T1)

FLEXORS OF THE FINGERS

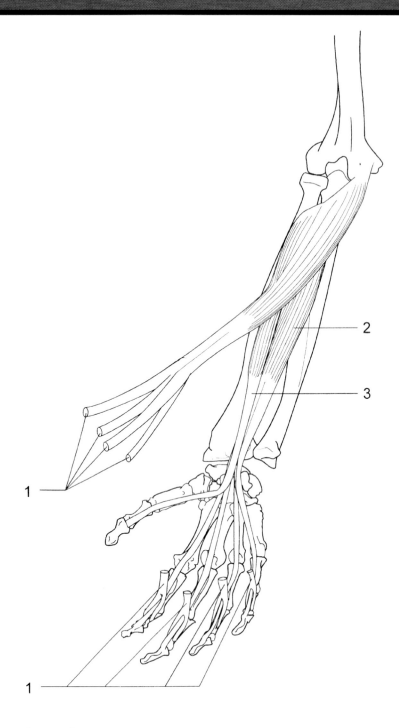

Forearm—anterior view

1. Flexor digitorum superficialis (cut)
2. Flexor digitorum profundus
3. Flexor pollicis longus

PRONATOR QUADRATUS

Forearm—anterior view

■ Origin
Anterior surface of distal part of shaft of ulna

■ Insertion
Lower portion of anterior surface of shaft of radius, distal part of lateral border of radius

■ Action
Pronates forearm and hand

■ Nerve
Anterior interosseous branch of median nerve (C8, T1)

■ Relationships
Deepest forearm muscle

BRACHIORADIALIS

Forearm—dorsal view

■ Origin
Upper two-thirds of lateral supracondylar ridge of humerus

■ Insertion
Base of styloid process and lateral surface of radius

■ Action
Flexes forearm

■ Nerve
Radial nerve (C5, C6)

EXTENSOR CARPI RADIALIS LONGUS

Forearm—dorsal view

■ **Origin**

Lower third of lateral supracondylar ridge of humerus

■ **Insertion**

Dorsal surface of the base of the second metacarpal bone

■ **Action**

Extends hand, synergist in abduction of hand with flexor carpi radialis

■ **Nerve**

Radial nerve (C6, C7)

EXTENSOR CARPI RADIALIS BREVIS

Forearm—dorsal view

■ **Origin**
Lateral epicondyle of humerus

■ **Insertion**
Dorsal surface of third metacarpal bone

■ **Action**
Extends hand, synergist in abduction of hand with flexor carpi radialis

■ **Nerve**
Radial nerve (C6, C7)

EXTENSOR DIGITORUM COMMUNIS

Forearm and hand—dorsal view

■ **Origin**
Common tendon attached to lateral epicondyle of humerus

■ **Insertion**
Lateral and dorsal surfaces of all the phalanges of the four fingers

■ **Action**
Extends the fingers and wrist

■ **Nerve**
Deep branch of radial nerve (C6–C8)

■ **Relationships**
Tends to hyperextend the metacarpophalangeal joint causing "claw hand"; its action is balanced by the lumbricales and interossei

EXTENSOR DIGITI MINIMI

Forearm and hand—dorsal view

■ **Origin**
Common tendon attached to lateral epicondyle of humerus

■ **Insertion**
Dorsal surface of base of first phalanx of fifth finger

■ **Action**
Extends fifth finger

■ **Nerve**
Radial nerve (C6–C8)

EXTENSOR CARPI ULNARIS

Forearm and hand—dorsal view

■ **Origin**
Common tendon attached to lateral epicondyle of humerus

■ **Insertion**
Dorsal surface of base of fifth metacarpal bone

■ **Action**
Extends hand, synergist in adduction of hand with flexor carpi ulnaris

■ **Nerve**
Radial nerve (C6–C8)

SUPINATOR

Forearm and hand—anterior view

■ Origin
Lateral epicondyle of humerus, lateral ligament (radial collateral) of elbow, annular ligament of superior radioulnar joint, supinator crest of ulna

■ Insertion
Dorsal and lateral surfaces of upper third of radius

■ Action
Supinates forearm

■ Nerve
Radial nerve (C6)

ABDUCTOR POLLICIS LONGUS

Forearm and hand—dorsal view

■ **Origin**
Posterior (dorsal) surface of shaft of radius, ulna, interosseous membrane

■ **Insertion**
Dorsal surface of base of first metacarpal bone

■ **Action**
Abducts, laterally rotates, and extends thumb; abducts wrist

■ **Nerve**
Radial nerve (C6, C7)

EXTENSOR POLLICIS BREVIS

Forearm and hand—dorsal view

■ **Origin**
Dorsal surface of radius, adjacent part of interosseous membrane

■ **Insertion**
Base of proximal phalanx of thumb

■ **Action**
Extends thumb, abducts hand

■ **Nerve**
Radial nerve (C6, C7)

EXTENSOR POLLICIS LONGUS

Forearm and hand—dorsal view

■ **Origin**
Middle third of dorsal surface of ulna, interosseous membrane

■ **Insertion**
Base of distal phalanx of thumb

■ **Action**
Extends thumb

■ **Nerve**
Radial nerve (C6–C8)

EXTENSORS OF THE THUMB

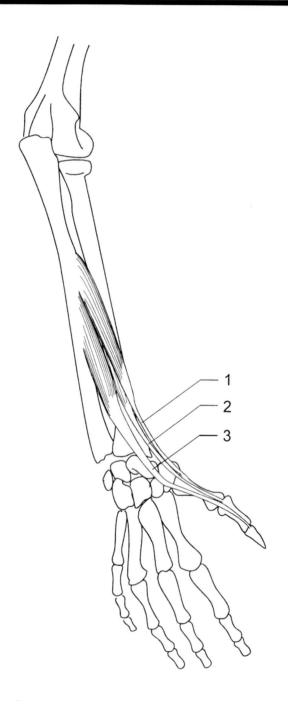

Forearm—posterior view

1. Abductor pollicis longus
2. Extensor pollicis brevis
3. Extensor pollicis longus

EXTENSOR INDICIS

Forearm and hand—dorsal view

■ **Origin**
Posterior surface of ulna and adjacent part of interosseous membrane

■ **Insertion**
Extensor expansion on dorsal surface of proximal phalanx of index finger

■ **Action**
Extends index finger

■ **Nerve**
Radial nerve (C6–C8)

PALMARIS BREVIS

Hand—palmar view

■ Origin
Flexor retinaculum, palmar aponeurosis

■ Insertion
Skin of the palm

■ Action
Corrugates skin of palm

■ Nerve
Ulnar nerve (C8)

ABDUCTOR POLLICIS BREVIS *(Thenar Eminence)*

Hand—palmar view

■ **Origin**
Tubercle of scaphoid, tubercle of trapezium, flexor retinaculum

■ **Insertion**
Base of proximal phalanx of thumb

■ **Action**
Abducts thumb and moves it anteriorly, acts together with other muscles of thenar eminence to oppose thumb to other fingers

■ **Nerve**
Median (C6, C7)

Note: The abductor pollicis brevis, flexor pollicis brevis, and opponens pollicis form the thenar eminence at the base of the thumb.

FLEXOR POLLICIS BREVIS *(Thenar Eminence)*

Hand—palmar view

■ **Origin**
Flexor retinaculum and trapezium, and first metacarpal bone

■ **Insertion**
Base of proximal phalanx of thumb

■ **Action**
Flexes metacarpophalangeal joint of thumb, assists in abduction and rotation of thumb, acts together with other muscles of thenar eminence to oppose thumb to other fingers

■ **Nerve**
Lateral portion—median nerve (C6, C7)

Medial portion—ulnar nerve (C8, T1)

OPPONENS POLLICIS *(Thenar Eminence)*

Hand—palmar view

■ **Origin**
Flexor retinaculum, tubercle of trapezium

■ **Insertion**
Lateral border of first metacarpal bone

■ **Action**
Rotates thumb into opposition with fingers, acts together with other muscles of thenar eminence to oppose thumb to other fingers

■ **Nerve**
Median nerve (C6, C7)

ADDUCTOR POLLICIS

Hand—palmar view

■ Origin
Oblique head—anterior surfaces of second and third metacarpals, capitate, trapezoid

Transverse head—anterior surface of third metacarpal bone

■ Insertion
Medial side of base of proximal phalanx of the thumb

■ Action
Adducts thumb

■ Nerve
Ulnar nerve (C8, T1)

ABDUCTOR DIGITI MINIMI *(Hypothenar Eminence)*

Hand—palmar view

■ **Origin**
Pisiform bone, tendon of flexor carpi ulnaris

■ **Insertion**
Medial side of base of proximal phalanx of fifth finger

■ **Action**
Abducts fifth finger

■ **Nerve**
Ulnar nerve (C8, T1)

Note: The hypothenar eminence is less prominent than the thenar eminence, and the fifth finger obviously cannot oppose the other digits.

FLEXOR DIGITI MINIMI BREVIS *(Hypothenar Eminence)*

Hand—palmar view

■ Origin
Anterior surface of flexor retinaculum, hook of hamate

■ Insertion
Medial side of base of proximal phalanx of fifth finger

■ Action
Flexes fifth finger at metacarpophalangeal joint

■ Nerve
Ulnar nerve (C8, T1)

OPPONENS DIGITI MINIMI *(Hypothenar Eminence)*

Hand—palmar view

■ **Origin**
Anterior surface of flexor retinaculum, hook of hamate

■ **Insertion**
Whole length of medial border of fifth metacarpal bone

■ **Action**
Rotates fifth metacarpal bone, draws fifth metacarpal bone forward, assists flexor digiti minimi brevis in flexing carpometacarpal joint of fifth finger

■ **Nerve**
Ulnar nerve (C8, T1)

LUMBRICALES* *(Four Muscles)*

Hand—palmar view

■ Origin
Tendons of flexor digitorum profundus in palm

■ Insertion
Lateral side of corresponding tendon of extensor digitorum on fingers

■ Action
Extend fingers at interphalangeal joints, weakly flex fingers at metacarpophalangeal joints

■ Nerve
Lateral lumbricals (first and second)—median nerve (C6, C7)

Medial lumbricals (third and fourth)—ulnar nerve (C8)

■ Relationships
Assist extensor digitorum communis in extending fingers without hyperextension at the metacarpophalangeal joints

*Associated with the tendons of flexor digitorum profundus.

PALMAR INTEROSSEI

Hand—palmar view

■ Origin
First—medial side of base of first metacarpal bone

Second, third, and fourth—anterior surfaces of second, fourth, and fifth metacarpal bones

■ Insertion
First—medial side of base of proximal phalanx of thumb

Second—medial side of base of proximal phalanx of index finger

Third and fourth—lateral side of proximal phalanges of ring finger and fifth finger

■ Action
Adduct fingers toward center of third finger at metacarpophalangeal joints, assist in flexion of fingers at metacarpophalangeal joints

■ Nerve
Ulnar nerve (C8, T1)

Note: The palmar interosseus of the thumb, called the palmar interosseus of Henle, is usually absent.

DORSAL INTEROSSEI

Hand—palmar view

■ Origin
By two heads from adjacent sides of first and second, second and third, third and fourth, and fourth and fifth metacarpal bones

■ Insertion
First—lateral side of base of proximal phalanx of index finger

Second—lateral side of base of proximal phalanx of middle finger

Third—medial side of base of proximal phalanx of middle finger

Fourth—medial side of base of proximal phalanx of ring finger

■ Action
Abduct fingers away from center of third finger at metacarpophalangeal joints, assist in flexion of fingers at metacarpophalangeal joints

■ Nerve
Ulnar nerve (C8, T1)

CHAPTER EIGHT

Muscles of the Hip and Thigh

PSOAS MAJOR *(Part of Iliopsoas)*

Lumbar region, hip, and thigh—anterior view

■ Origin
Bases of transverse processes of all lumbar vertebrae, bodies of twelfth thoracic and all lumbar vertebrae, intervertebral disks above each lumbar vertebra

■ Insertion
Lesser trochanter of femur

■ Action
Flexes thigh at hip joint, flexes vertebral column

■ Nerve
Branches from lumbar plexus (L2, L3) and sometimes L1 or L4

Note: Some upper fibers insert onto the hip bone from the arcuate line to the iliopectineal eminence to form the *psoas minor.* This muscle has little function and is frequently absent.

ILIACUS *(Part of Iliopsoas)*

Lumbar region, hip, and thigh—anterior view

■ Origin
Upper two-thirds of iliac fossa, ala of sacrum and adjacent ligaments, anterior inferior iliac spine

■ Insertion
Onto tendon of psoas major, which continues into lesser trochanter of femur (together the two muscles form the iliopsoas)

■ Action
Flexes thigh at hip joint

■ Nerve
Femoral nerve (L2, L3)

Note: The iliacus brings swinging leg forward in walking or running.

PIRIFORMIS

Hip and thigh—anterior view

■ **Origin**
Internal surface of sacrum, sacrotuberous ligament

■ **Insertion**
Upper border of greater trochanter

■ **Action**
Laterally rotates thigh at hip joint, abducts thigh

■ **Nerve**
Anterior rami of first and second sacral nerves

Note: The common peroneal part of the sciatic nerve may emerge through the belly of the piriformis instead of below its inferior border along with the tibial part.

OBTURATOR INTERNUS

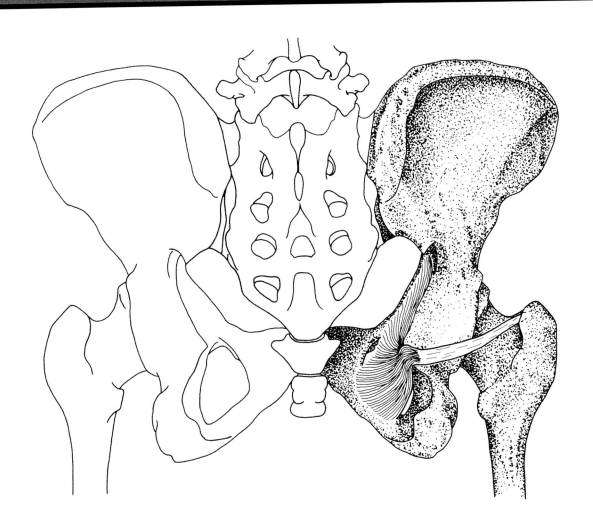

Hip—posterior view

■ Origin
Pelvic surface of obturator membrane and surrounding bones (ilium, ischium, pubis)

■ Insertion
Common tendon with superior and inferior gemelli to medial surface of greater trochanter

■ Action
Laterally rotates thigh at hip joint

■ Nerve
Nerve from sacral plexus (L5, S1–S3)

GEMELLUS SUPERIOR

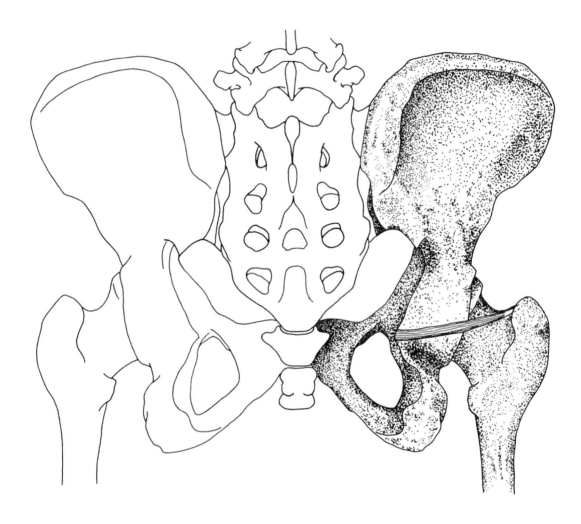

Hip—posterior view

■ **Origin**
Spine of ischium

■ **Insertion**
With tendon of obturator internus into upper border of greater trochanter

■ **Action**
Laterally rotates thigh at hip joint

■ **Nerve**
Branch of nerve to obturator internus from sacral plexus

GEMELLUS INFERIOR

Hip—posterior view

■ **Origin**

Upper margin of ischial tuberosity

■ **Insertion**

With tendon of obturator internus into upper border of greater trochanter

■ **Action**

Laterally rotates thigh at hip joint

■ **Nerve**

Branch of nerve to quadratus femoris from sacral plexus

OBTURATOR EXTERNUS

Hip and thigh—anterior view

■ **Origin**
Outer surface of superior and inferior rami of pubis and ramus of ischium surrounding obturator foramen

■ **Insertion**
Trochanteric fossa of femur

■ **Action**
Laterally rotates thigh

■ **Nerve**
Obturator nerve (L3, L4)

Note: Part of this muscle can be seen posteriorly by separating the gemellus inferior and quadratus femoris. It is deep within this cleft.

QUADRATUS FEMORIS

Hip and thigh—posterior view

■ **Origin**
Lateral border of ischial tuberosity

■ **Insertion**
Below intertrochanteric crest (quadrate line)

■ **Action**
Laterally rotates thigh at hip joint

■ **Nerve**
Branch from sacral plexus (L5, S1)

GLUTEUS MAXIMUS

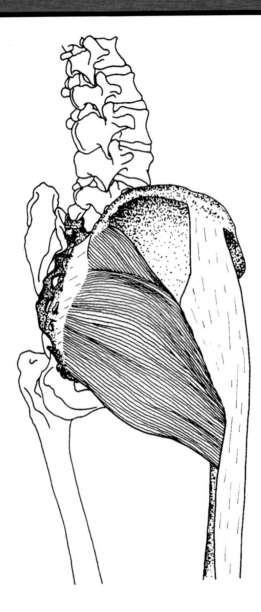

Hip and thigh—lateral view

■ Origin
Outer surface of ilium behind posterior gluteal line, adjacent posterior surface of sacrum and coccyx, sacrotuberous ligament, aponeurosis of erector spinae (sacrospinalis)

■ Insertion
Iliotibial tract of fascia lata, gluteal tuberosity of femur

■ Action
Upper part—abducts, laterally rotates thigh

Lower part—extends, laterally rotates thigh, extends trunk, assists in adduction of thigh

■ Nerve
Inferior gluteal nerve (L5, S1, S2)

Note: This is not a postural muscle; it is not used in walking but only in forceful extension, as in running, climbing, or rising from a seated position.

GLUTEUS MEDIUS

Hip and thigh—lateral view

■ Origin
Outer surface of ilium inferior to iliac crest

■ Insertion
Lateral surface of greater trochanter

■ Action
Abducts femur at hip joint and rotates thigh medially

■ Nerve
Superior gluteal nerve (L4, L5, S1)

Note: In locomotion, this muscle (along with the gluteus minimus) prevents the pelvis from dropping (adduction of thigh) toward the opposite swinging leg.

GLUTEUS MINIMUS

Hip and thigh—lateral view

■ Origin
Outer surface of ilium between middle (anterior) and inferior gluteal lines

■ Insertion
Anterior surface of greater trochanter

■ Action
Abducts femur at hip joint and rotates thigh medially

■ Nerve
Superior gluteal nerve (L4, L5, S1)

Note: See note on gluteus medius.

MUSCLES OF THE HIP

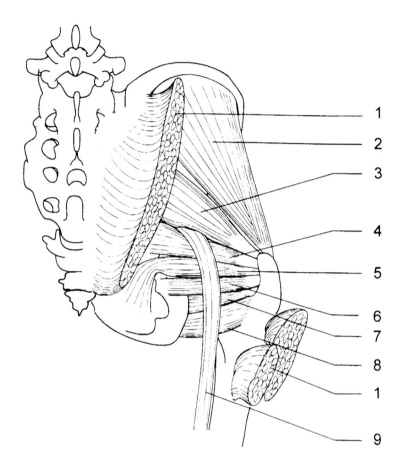

Hip—posterior view

1. Gluteus maximus (cut)
2. Gluteus medius
3. Piriformis
4. Gemellus superior
5. Obturator internus

6. Gemellus inferior
7. Obturator externus
8. Quadratus femoris
9. Sciatic nerve

Note: Gemellus inferior and quadratus femoris have been shown separated to expose the deeply placed obturator externus.

TENSOR FASCIAE LATAE

Hip and thigh—lateral view

■ Origin
Outer edge of iliac crest between anterior superior iliac spine and iliac tubercle

■ Insertion
Iliotibial tract on upper part of thigh

■ Action
Flexes, abducts thigh

■ Nerve
Superior gluteal nerve (L4, L5, S1)

Note: This muscle, along with gluteus maximus, draws the fascia lata upward, stabilizing the knee.

SARTORIUS

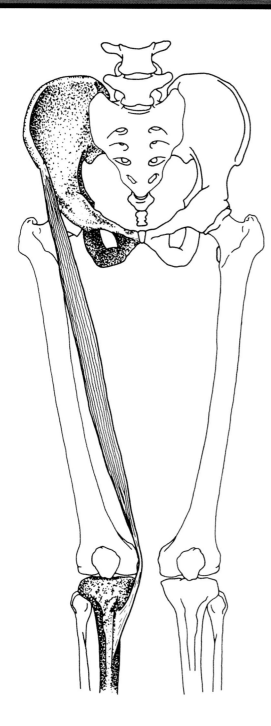

Hip, thigh, and leg—anterior view

■ Origin
Anterior superior iliac spine and area immediately below it

■ Insertion
Upper part of medial surface of shaft of tibia

■ Action
Flexes, abducts, and laterally rotates thigh at hip joint, flexes and slightly medially rotates leg at knee joint after flexion

■ Nerve
Femoral nerve (L2, L3)

■ Relationships
Insertions of sartorius, gracilis, and semitendinosus fuse on the medial tibia; these tendons, called the pes anserinus (goose foot), give medial support to the knee

Note: This muscle is used to bring swinging leg forward in walking and running.

RECTUS FEMORIS   *(One of quadriceps femoris)*

Hip, thigh, and leg—anterior view

■ Origin
Anterior head—anterior inferior iliac spine

Posterior head—ilium above acetabulum

■ Insertion
Patella, then by patellar ligament to tuberosity of the tibia

■ Action
Extends leg at knee joint, flexes thigh at hip joint

■ Nerve
Femoral nerve (L2–L4)

Note: This muscle is used when thigh flexion and leg extension are needed together, such as in kicking a football. In walking, the quadriceps prevent the knee from flexing during heel strike and early support phase.

VASTUS LATERALIS *(One of quadriceps femoris)*

Hip, thigh, and leg—anterior view

■ Origin
Intertrochanteric line, inferior border of greater trochanter, gluteal tuberosity, lateral lip of linea aspera of femur

■ Insertion
Lateral margin of patella, then by patellar ligament to tuberosity of tibia

■ Action
Extends leg at knee joint

■ Nerve
Femoral nerve (L2–L4)

VASTUS MEDIALIS *(One of quadriceps femoris)*

Hip, thigh, and leg—anterior view

■ Origin
Intertrochanteric line, medial lip of linea aspera of femur, medial intermuscular septum, medial supracondylar line

■ Insertion
Medial border of the patella, then by patellar ligament into tibial tuberosity, medial condyle of tibia

■ Action
Extends leg at knee joint

■ Nerve
Femoral nerve (L2–L4)

VASTUS INTERMEDIUS *(One of quadriceps femoris)*

Hip, thigh, and leg—anterior view

■ Origin
Anterior and lateral surfaces of upper two-thirds of femur, lateral intermuscular septum, linea aspera, lateral supracondylar line

■ Insertion
Deep aspect of quadriceps tendon, then through patella to tibial tuberosity

■ Action
Extends leg at knee joint

■ Nerve
Femoral nerve (L2–L4)

Note: A few bundles of fibers from this muscle insert onto the upper part of the joint capsule of the knee. They probably draw the capsule superiorly during extension of the leg, preventing it from binding in the joint. They are called *articularis genus*.

MUSCLES OF THE ANTERIOR THIGH

Hip and thigh—anterior view

1. Tensor fasciae latae
2. Iliotibial tract
3. Vastus lateralis (quadriceps femoris)
4. Vastus intermedius (quadriceps femoris)
5. Rectus femoris (cut) (quadriceps femoris)
6. Sartorius
7. Vastus medialis (quadriceps femoris)

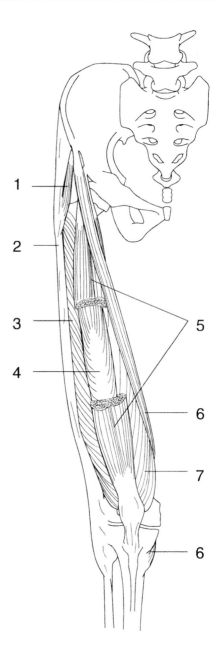

BICEPS FEMORIS *(Part of Hamstrings)*

Hip and thigh—posterior view

■ Origin
Long head—ischial tuberosity, sacrotuberous ligament

Short head—linea aspera, lateral supracondylar ridge, lateral intermuscular septum

■ Insertion
Lateral side of head of fibula and lateral condyle of tibia

■ Action
Flexes leg at knee joint, long head also extends thigh at hip joint

■ Nerve
Long head—tibial part of sciatic nerve (S1–S3)

Short head—common peroneal part of sciatic nerve (L5, S1, S2)

Note: During walking or running, the hamstrings are used to slow down the leg at the end of its swing and prevent the trunk from flexing at the hip. They are susceptible to being strained by resisting the momentum of these body parts.

SEMITENDINOSUS *(Part of Hamstrings)*

Hip and thigh—posterior view

■ Origin
Ischial tuberosity

■ Insertion
Medial surface of shaft of tibia

■ Action
Flexes and slightly medially rotates leg at knee joint after flexion, extends thigh at hip joint

■ Nerve
Tibial portion of sciatic nerve (L5, S1, S2)

Note: See note on biceps femoris and Relationships section on sartorius.

SEMIMEMBRANOSUS *(Part of Hamstrings)*

Hip and thigh—posterior view

■ Origin
Ischial tuberosity

■ Insertion
Posterior part of medial condyle of tibia

■ Action
Flexes and slightly medially rotates leg at knee joint after flexion, extends thigh at hip joint

■ Nerve
Tibial portion of sciatic nerve (L5, S1, S2)

Note: See note on biceps femoris.

HAMSTRING MUSCLES

Hip and thigh—posterior view

1. Sciatic nerve
2. Quadratus femoris
3. Biceps femoris
4. Semimembranosus
5. Semitendinosus
6. Tibial nerve
7. Common peroneal nerve

Note: The common peroneal nerve is exposed to compression and damage as it passes over the head of the fibula. The quadratus femoris, a lateral rotator, is included for reference.

GRACILIS

Hip and thigh—anterior view

■ Origin
Lower margin of body and inferior ramus of pubis

■ Insertion
Upper part of medial surface of shaft of tibia

■ Action
Adducts thigh at hip joint and flexes leg at knee joint; with leg flexed, assists in medial rotation

■ Nerve
Obturator nerve (L3, L4)

Note: See Relationships section on sartorius.

PECTINEUS

Hip and thigh—anterior view

■ Origin
Pectineal line on superior ramus of pubis

■ Insertion
From lesser trochanter to linea aspera of femur

■ Action
Flexes thigh, assists in adduction when hip is flexed

■ Nerve
Femoral nerve (L2–L4), (sometimes a branch of obturator nerve)

Note: The rotating function of this and other hip muscles is controversial and probably depends on whether the hip is flexed or extended and adducted or abducted.

ADDUCTOR LONGUS

Hip and thigh—posterior view

■ **Origin**
Anterior of body of pubis

■ **Insertion**
Medial lip of linea aspera

■ **Action**
Adducts, flexes thigh, assists in medial rotation

■ **Nerve**
Obturator nerve (L3, L4)

ADDUCTOR BREVIS

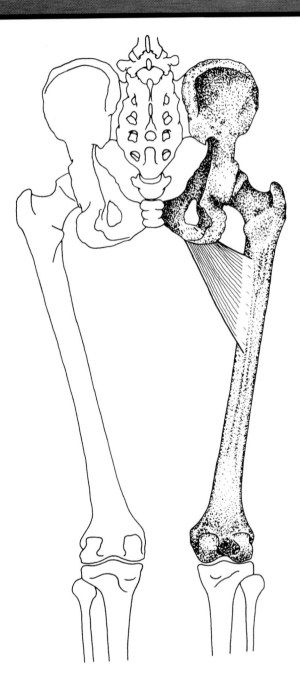

Hip and thigh—posterior view

■ Origin
Outer surface of inferior ramus of pubis

■ Insertion
From below lesser trochanter to linea aspera and into proximal part of linea aspera

■ Action
Adducts thigh, assists in flexion, medial rotation

■ Nerve
Obturator nerve (L3, L4)

ADDUCTOR MAGNUS

Hip and thigh—posterior view

■ Origin
Inferior ramus of pubis, and ramus and lower part of tuberosity of ischium

■ Insertion
Linea aspera, adductor tubercle of femur

■ Action
Adducts, extends thigh, lower portion (adductor tubercle insertion) assists in medial rotation

■ Nerve
Obturator nerve (L3, L4), sciatic nerve

Note: The linea aspera insertion may assist in lateral rotation.

HIP FLEXORS AND ADDUCTORS

Hip and thigh—anterior view

1. Psoas major
2. Iliacus
3. Inguinal ligament
4. Femoral nerve, vein, artery
5. Pectineus
6. Adductor brevis
7. Adductor longus (cut)
8. Adductor magnus
9. Gracilis

CHAPTER NINE

Muscles of the Leg and Foot

TIBIALIS ANTERIOR

Leg—anterolateral view

■ Origin
Lateral condyle of tibia, upper half of lateral surface of tibia, interosseous membrane

■ Insertion
Medial side and plantar surface of medial cuneiform bone, and base of first metatarsal bone

■ Action
Dorsiflexes foot at ankle joint, inverts (supinates) foot

■ Nerve
Deep peroneal nerve (L4, L5, S1)

EXTENSOR HALLUCIS LONGUS

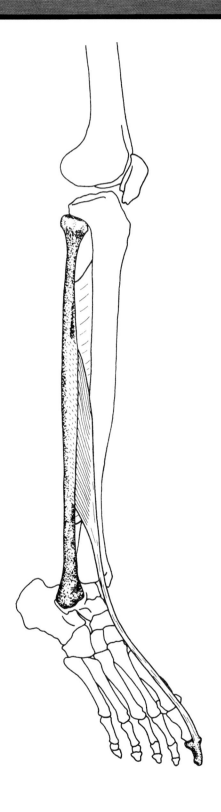

Leg—anterolateral view

■ Origin
Middle half of anterior surface of fibula and interosseous membrane

■ Insertion
Base of distal phalanx of great toe

■ Action
Extends, hyperextends great toe, dorsiflexes and inverts (supinates) foot

■ Nerve
Deep peroneal nerve (L4, L5, S1)

EXTENSOR DIGITORUM LONGUS

Foot—anterolateral view

■ Origin
Upper two-thirds of anterior surface of fibula, interosseous membrane, lateral condyle of tibia

■ Insertion
Along dorsal surface of four lateral toes, and then to bases of middle and distal phalanges

■ Action
Extends toes, dorsiflexes foot at ankle, everts foot

■ Nerve
Deep peroneal nerve (L4, L5, S1)

Note: The lower lateral part of this muscle makes a separate insertion onto the dorsal surface of the fifth metatarsal and is called *peroneus tertius*.

Leg—anterolateral view

PERONEUS TERTIUS *(Lower lateral part of extensor digitorum longus)*

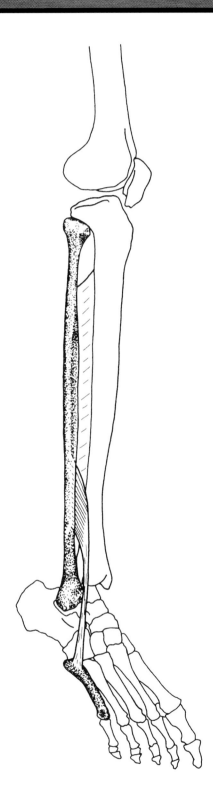

Leg—anterolateral view

■ Origin
Lower third of anterior surface of fibula and interosseous membrane

■ Insertion
Dorsal surface of base of fifth metatarsal bone

■ Action
Dorsiflexes and everts foot

■ Nerve
Deep peroneal nerve (L4, L5, S1)

ANTERIOR AND LATERAL LEG MUSCLES

Leg—anterolateral view

1. Peroneus longus
2. Peroneus brevis
3. Peroneus tertius
4. Tibialis anterior
5. Extensor retinaculum
6. Extensor hallucis longus
7. Extensor digitorum longus
8. Extensor digitorum brevis

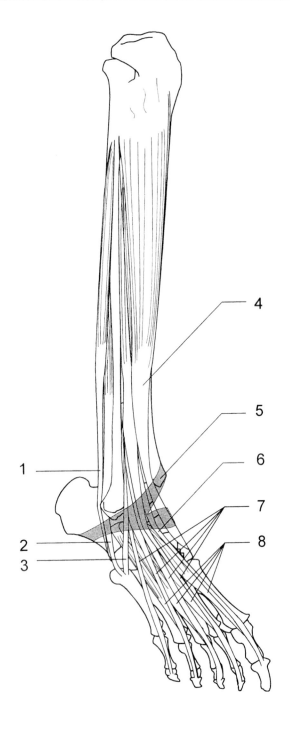

GASTROCNEMIUS *(Part of triceps surae)*

Leg—posterior view

■ Origin
Lateral head—lateral condyle and posterior surface of femur

Medial head—popliteal surface of femur above medial condyle

■ Insertion
Posterior surface of the calcaneus

■ Action
Plantar flexes foot, flexes leg at knee

■ Nerve
Tibial nerve (S1, S2)

SOLEUS *(Part of triceps surae)*

Leg—posterior view

■ Origin
Posterior surface of the tibia (soleal line), upper third of posterior surface of fibula, fibrous arch between tibia and fibula

■ Insertion
Posterior surface of the calcaneus

■ Action
Plantar flexes foot

■ Nerve
Tibial nerve (S1, S2)

PLANTARIS

Leg—posterior view

■ **Origin**
Lateral supracondylar ridge of femur, oblique popliteal ligament

■ **Insertion**
Posterior surface of the calcaneus

■ **Action**
Plantar flexes foot, flexes leg

■ **Nerve**
Tibial nerve (L4, L5, S1)

MUSCLES OF THE CALF

Leg—posterior view

1. Soleus
2. Plantaris
3. Gastrocnemius (cut)

POPLITEUS

Leg—posterior view

■ Origin
Lateral surface of lateral condyle of femur

■ Insertion
Upper part of posterior surface of tibia

■ Action
Rotates leg medially, flexes leg

■ Nerve
Tibial nerve (L4, L5, S1)

Note: Stern contends that this muscle stabilizes the knee by preventing lateral rotation of the tibia during medial rotation of the thigh while the foot is planted.

Reference: Stern, J. T. *Essentials of Gross Anatomy*, F. A. Davis Company, Philadelphia, 1988.

FLEXOR HALLUCIS LONGUS

Leg—posterior view

■ Origin
Lower two-thirds of posterior surface of shaft of fibula, posterior intermuscular septum, interosseous membrane

■ Insertion
Base of distal phalanx of great toe

■ Action
Flexes distal phalanx of great toe, assists in plantar flexing foot, inverts foot

■ Nerve
Tibial nerve (L5, S1, S2)

Note: This muscle is important in pushing off the surface in walking, running, jumping.

FLEXOR DIGITORUM LONGUS

Foot—plantar view

■ Origin
Medial part of posterior surface of tibia

■ Insertion
Bases of distal phalanges of second, third, fourth, and fifth toes

■ Action
Flexes distal phalanges of lateral four toes, assists in plantar flexing foot, inverts foot

■ Nerve
Tibial nerve (L5, S1)

Leg—posterior view

TIBIALIS POSTERIOR

Foot—plantar view

■ Origin
Lateral part of posterior surface of tibia, interosseous membrane, proximal half of posterior surface of fibula

■ Insertion
Tuberosity of navicular bone, cuboid, cuneiforms, second, third, and fourth metatarsals, sustentaculum tali of calcaneus

■ Action
Plantar flexes, inverts foot

■ Nerve
Tibial nerve (L5, S1)

Leg—posterior view

DEEP POSTERIOR LEG MUSCLES

Leg—posterior view

1. Popliteus
2. Tibialis posterior
2a. Tendon of tibialis posterior
3. Flexor hallucis longus (cut)
4. Flexor digitorum longus (cut)
5. Medial malleolus

PERONEUS LONGUS

Foot—plantar view

■ **Origin**
Upper two-thirds of lateral surface of fibula

■ **Insertion**
Lateral side of medial cuneiform, base of first metatarsal

■ **Action**
Plantar flexes, everts foot

■ **Nerve**
Superficial peroneal nerve (L4, L5, S1)

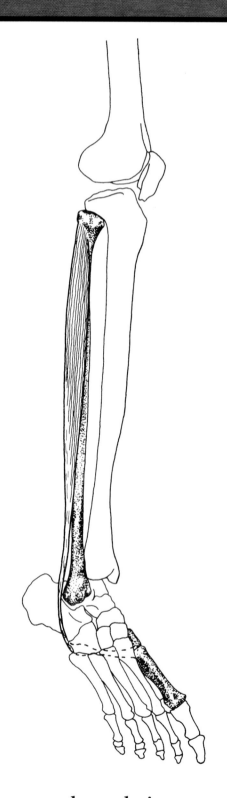

Leg—anterolateral view

PERONEUS BREVIS

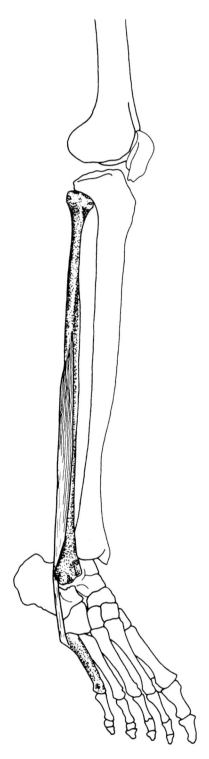

Leg—anterolateral view

■ **Origin**
Lower two-thirds of lateral surface of fibula

■ **Insertion**
Lateral side of base of fifth metatarsal bone

■ **Action**
Everts, plantar flexes foot

■ **Nerve**
Superficial peroneal nerve (L4, L5, S1)

EXTENSOR DIGITORUM BREVIS

Foot—anterolateral view

■ **Origin**
Anterior and lateral surfaces of calcaneus, lateral talocalcaneal ligament, inferior extensor retinaculum

■ **Insertion**
Into base of proximal phalanx of great toe; into lateral sides of tendons of extensor digitorum longus of second, third, and fourth toes

■ **Action**
Extends the four toes

■ **Nerve**
Deep peroneal nerve (L5, S1)

ABDUCTOR HALLUCIS *(First Layer)*

Foot—plantar view

■ Origin
Tuberosity of calcaneus, flexor retinaculum, plantar aponeurosis

■ Insertion
Medial side of base of proximal phalanx of great toe

■ Action
Stabilizes great toe (with adductor hallucis)

■ Nerve
Medial plantar nerve (L4, L5)

Note: The muscles of the sole of the foot can be divided into four layers (from superficial to deep):

First layer—abductor hallucis, flexor digitorum brevis, abductor digiti minimi

Second layer—quadratus plantae, lumbricales (tendons of flexor hallucis longus and flexor digitorum longus pass through this layer)

Third layer—flexor hallucis brevis, adductor hallucis, flexor digiti minimi brevis

Fourth layer—interossei (tendons of tibialis posterior and peroneus longus pass through this layer)

FLEXOR DIGITORUM BREVIS *(First Layer)*

Foot—plantar view

■ **Origin**
Tuberosity of calcaneus, plantar aponeurosis

■ **Insertion**
Sides of middle phalanges of second to fifth toes

■ **Action**
Flexes proximal phalanges and extends distal phalanges of second through fifth toes

■ **Nerve**
Medial plantar nerve (L4, L5)

ABDUCTOR DIGITI MINIMI *(First Layer)*

Foot—plantar view

■ **Origin**
Tuberosity of calcaneus, plantar aponeurosis

■ **Insertion**
Lateral side of proximal phalanx of fifth toe

■ **Action**
Abducts fifth toe

■ **Nerve**
Lateral plantar nerve (S1, S2)

QUADRATUS PLANTAE *(Second Layer)*

Foot—plantar view

■ Origin
Medial head—medial surface of calcaneus

Lateral head—lateral border of inferior surface of calcaneus

■ Insertion
Lateral margin of tendon of flexor digitorum longus

■ Action
Flexes terminal phalanges of second through fifth toes

■ Nerve
Lateral plantar nerve (S1, S2)

LUMBRICALES *(Second Layer)*

Foot—plantar view

■ Origin
Tendons of flexor digitorum longus

■ Insertion
Dorsal surfaces of proximal phalanges

■ Action
Flex proximal phalanges of second through fifth toes

■ Nerve
First lumbricalis—medial plantar nerve (L4, L5)

Second through fifth lumbricales—lateral plantar nerve (S1, S2)

FLEXOR HALLUCIS BREVIS *(Third Layer)*

Foot—plantar view

■ Origin
Cuboid bone, lateral cuneiform bone

■ Insertion
Medial part—medial side of base of proximal phalanx of great toe

Lateral part—lateral side of base of proximal phalanx of great toe

■ Action
Flexes proximal phalanx of great toe

■ Nerve
Medial plantar nerve (L4, L5, S1)

ADDUCTOR HALLUCIS *(Third Layer)*

Foot—plantar view

■ Origin

Oblique head—second, third, and fourth metatarsal bones, and sheath of peroneus longus tendon

Transverse head—plantar metatarsophalangeal ligaments of third, fourth, and fifth toes, and transverse metatarsal ligaments

■ Insertion

Lateral side of base of proximal phalanx of great toe

■ Action

Stabilizes great toe (with abductor hallucis)

■ Nerve

Lateral plantar nerve (S1, S2)

FLEXOR DIGITI MINIMI BREVIS *(Third Layer)*

Foot—plantar view

■ **Origin**
Base of fifth metatarsal, sheath of peroneus longus tendon

■ **Insertion**
Lateral side of base of proximal phalanx of fifth toe

■ **Action**
Flexes proximal phalanx of fifth toe

■ **Nerve**
Lateral plantar nerve (S1, S2)

DORSAL INTEROSSEI *(Fourth Layer; Four Muscles)*

Foot—anterolateral view

■ Origin
Adjacent sides of metatarsal bones

■ Insertion
Bases of proximal phalanges

First—medial side of proximal phalanx of second toe

Second, third, fourth—lateral sides of proximal phalanges of second, third, and fourth toes

■ Action
Abduct toes, flex proximal phalanges

■ Nerve
Lateral plantar nerve (S1, S2)

PLANTAR INTEROSSEI *(Fourth Layer; Three Muscles)*

Foot—plantar view

■ Origin
Bases and medial sides of third, fourth, and fifth metatarsal bones

■ Insertion
Medial sides of bases of proximal phalanges of same toes

■ Action
Adduct toes, flex proximal phalanges

■ Nerve
Lateral plantar nerve (S1, S2)

Alphabetical Listing of Muscles

Index

Notes

Notes

Notes

Notes

Notes

Notes

Notes

Notes

Notes

Notes

Notes

Notes

Notes